THE RANDOM UNIVERSE

The Random Universe

How Models and Probability Help Us Make Sense of the Cosmos

ANDREW H. JAFFE

Yale

UNIVERSITY PRESS

NEW HAVEN AND LONDON

Published with assistance from the Louis Stern Memorial Fund.

Illustrations are by the author unless otherwise credited.

Yale University Press books may be purchased in quantity for educational, business,
or promotional use. For information, please e-mail sales.press@yale.edu (U.S. office)
or sales@yaleup.co.uk (U.K. office).

Set in Electra type by IDS Infotech, Ltd.

Printed and bound by CPI Group (UK) Ltd, Croydon, CR0 4YY

Library of Congress Control Number: 2025933303
ISBN 978-0-300-25050-3 (hardcover)

A catalogue record for this book is available from the British Library.

Authorized Representative in the EU:
Easy Access System Europe, Mustamäe tee 50,
10621 Tallinn, Estonia, gpsr.requests@easproject.com

10 9 8 7 6 5 4 3 2 1

For Lisa, Tallulah, and Olive

CONTENTS

Part II. The Randomness of the Universe

Part III. Cosmology in a Random Universe

In a *New York Times* opinion article, David Brooks tried to explain or understand his surprise at the outcome of the 2024 US elections: "We all walk around with mental models of reality in our heads. Our mental models help us make sense of the buzzing, blooming confusion of the world. Our mental models help us anticipate what's about to happen. Our mental models guide us as we make decisions about how to get the results we want. . . . Where did we get our current models? Well, we get models from our experience, our peers, the educational system, the media and popular culture." But these kinds of models are also how scientists interpret our observations and the results of our experiments, refining and combining them into theories of how the world works. You can't understand the science unless you understand the models and how they were built. This book is about these "mental models of reality," from an infant's first sense of cause and effect to modern scientists' theories of the Big Bang.

But the models, and the data that contribute to them, are always incomplete and are often contradictory: we are always learning and acting under conditions of uncertainty. Is all the information equally valid? Are all models—all interpretations of the data—equally likely? Adjudicating among them requires a mathematical tool, probability, even if we don't always use it consciously.

These are the questions that have driven me throughout my life and career, and this book is my attempt to answer them for a broader audience, outside the confines of a scientific journal.

We will have many guides along the way to understanding these models and the accompanying mathematics. In the first part of the book, I will try to encourage this way of understanding the world by holding some long-distance discussions with philosophers from David Hume in the Enlightenment to more recent thinkers like Karl Popper and Imre Lakatos. Along the way, we'll also hear from Isaac Newton, Pierre-Simon Laplace, and twentieth-century physicists like Albert Einstein and Arthur Eddington. The Reverend Thomas Bayes will hover over much of our discussion, showing us how and why we will need the mathematics of probability.

In the second part, we will look at the different areas of math and physics that require a probabilistic understanding. We will try to understand the meaning of randomness and how it can arise even in physical theories. We will touch on the concepts of entropy and information, which are needed to understand everything from artificial intelligence to how the behavior of matter is tied to its properties on the very smallest scales. And we will discuss quantum mechanics, a beautiful and still sometimes mysterious theory even a century after its development that can only be written down in terms of probabilities, seemingly part of the very fabric of the Universe.

And in the last part of the book, we will combine these physical theories to give us a picture of modern cosmology, a model for how the Universe has expanded and evolved since the Big Bang almost 14 billion years ago. We will use these models to help us interpret data from the Planck satellite (which I had the privilege of working on for almost two decades of my scientific career) and try to understand how we might be able to further refine this model and what might get in our way.

The test of your own understanding is how well you can explain your model of the Universe. *The Random Universe* is my attempt to hold myself to account in just this way. It's intended for everyone who has wondered about the Universe and wondered how we can even start answering questions on such a grand scale. Some topics are only covered in brief, but the book should be self-contained, although I am sure that readers with more exposure to physics and cosmology at the level of popular magazines and websites will find some topics familiar. I have mostly eschewed formulas, although a few are needed for a clear understanding of probability (and a few more are so beautiful that it would be a shame not to at least write them down here). And I hope that the book raises as many questions as it answers, so we can all keep building models of our world.

This book would not exist without the support, patience, and love of my family: Lisa, my wife, partner, and coconspirator, indispensable to the creation of all my models as well as this book, along with my daughters Tallulah and Olive, who force me to confront those models whenever they can. I would also like to thank my parents, Lynne and Ed Jaffe, for many years of encouragement, and my collaborators and interlocutors throughout my career, for both reinforcing and challenging my understanding of probability and model-building, especially Josh Frieman, Michael Turner, Marc Kamionkowski, Nick Kaiser, Dick Bond, Glenn Starkman, Daniel Mortlock, Alan Heavens, and Elena Sellentin. Finally, I must thank Pedro Ferreira, who inspired me with the phrase that eventually became the title for this book.

PART ONE

How We Learn About the World

ONE

We Are Scientists

We are all scientists.

As soon as we open our eyes each morning, we are confronted with information we have to make sense of. Our brains process the signals from the retinas at the backs of our eyes, the vibrations of our eardrums, and the feeling of our bed beneath us, the blanket on top. We reach out when we hear the alarm clock and perform an experiment, trying to find the button. At the same time, our feelings of hunger, of being about to sneeze, of an itch on our elbow, let us build a mental model of ourselves as part of that world. Here, too, we plan our experiments: eating breakfast should assuage that feeling of hunger; scratching that itch should make it go away; coffee will jolt us awake.

Without these models we could not comprehend the world around us. We could not even interpret the data from our senses. We have built these models over our years of experience and thought to inform our beliefs and guide our actions. Some may be innate, encoded by the process of evolution to make us fit for our environment; others we may have learned early on from our parents and teachers; some are constantly refined and tested by our need to understand the changing world around us. Our actions in the world are rough experiments designed by us to improve our models and our ability to use those models to get what we want.

The work of building models and using them to understand the world around us is exactly the job of us scientists. Our models are scientific hypotheses and theories, and our actions are experiments to help us understand, prove, or disprove those theories.

The sun is a glowing ball of gas, tens of millions of miles away, lit by nuclear reactions occurring deep in its interior. The stars in the night sky are all nuclear furnaces like our own sun, but even the nearest of them are millions of times farther away, appearing to us as single points of light.

These stars are arranged into agglomerations of tens of billions called galaxies, and we have seen galaxies so far away that their light comes to us from when the Universe was less than one percent of its current age.

These galaxies are all moving away from each other, a sign of the expansion of the Universe as a whole, and have been since about 14 billion years ago when everything that is now visible to our eyes and telescopes—all those stars, and the planets around them, and the galaxies they make up, everything—was compressed into a volume smaller than a pea.

How can we possibly know all this?

No human has ever been farther than the moon, less than one-quarter of one percent of the distance to the sun, whose interior properties we claim to know so well. Nor have scientists been able to peer inside any star, much less collect and analyze samples of their interiors to perform experiments on them. Even our most powerful telescopes cannot pick out the light from individual stars in most of the millions of galaxies that we've managed to observe.

Given these limitations, how can we know what's going on inside the sun or any other star? How do we know that those points of light in the night sky are stars like our sun? How can we possibly know how old the Universe is if we weren't present at the moment of its creation?

One of my kids' books asks, "If the dinosaurs went extinct millions of years ago, how do we know they exist"?[1] We can't see them walk the Earth, but we claim to know that they did, over a period from about 250 million to 60 million years ago. The first humans didn't appear until less than half a million years ago, and the first remains of dinosaurs were only really identified as such in the nineteenth century. How can we have any precise knowledge of these creatures from so long ago?

These kinds of questions about our knowledge of things we have not directly experienced don't just apply to these grand scientific facts about the

Universe. We grapple with them every day. How do I know when my bus will arrive, and whether it will get me to work on time, when I don't know the details of today's London traffic? How do I know that the baby crying in the other room wants her bottle? How do I know if it's still dark outside when I wake up? How do I know how my colleagues at work will react when I ask them to schedule a meeting? How can I tell the song on the radio is by the Beatles if I've never heard it before?[2]

What does it mean to *know* something that I've never observed, or even can't observe, directly? If I stay on this skeptical path, how can I be sure of anything at all?

Well, I know what is going on inside a star because scientists over the last few hundred years have developed a set of interlocking theories for gravity, the force that holds a star together; for the reactions that release energy through the process of nuclear fusion, when gravity squeezes the center of the star tight enough; and for the way that energy interacts with the rest of the matter in the star, allowing it to make its way to the surface of the star and eventually to our eyes and telescopes.

That is, we have a *model* for the sun, which we have learned we can also apply to all the other stars that we can see. We have tested this model against our observations of the sun and those other stars. And those models rely on yet more models, of gravity and nuclear reactions and atoms and the transport of heat and light, which we can test in more convenient laboratory locations here on Earth: it all comes back to experiments and observations with knobs and dials and screens—and the bodies and senses and brains of human scientists.

The sun and the stars and the edge of the Universe are inaccessible, but no more so than the interior lives of other people. When you are an adolescent, much of the work you do seems to be building a model that helps you answer the question "Do other people like me?" Some of this model gets built up by experimentation—trying to make friends, changing the clothes you wear, figuring out what music to listen to and sports to play. Like pretty much everyone else, my experiments were sometimes successful, for example when I ran for student government in eighth grade on a lark and managed to get elected. I guess I had a good enough model for my classmates that year, although I'm also sure that I could not have explained it very well. What, exactly, did they like about me?

Of course, sometimes the models were less successful—like when I tried to ask my middle school crush on a date that same year. I'm not sure I really expected that experiment to work, but I certainly hoped it would.

But building models of the world, and especially of the other people in it, doesn't end after childhood. We still strive to be liked, and some people's actions remain inexplicable despite the social theories that we've built up over the years. This becomes more formalized when we apply these theories to help get a new job: what we put in our résumé is a codification of how we have modeled the hiring process. And the hiring panel has a model, too, of the job that they are trying to fill, and of the traits that they want for the person in that position. It all gets meta: the applicant changes her behavior to match her model of the panel's model of the new job and tries to convince them that their model of her is the right match for their own model of the job.

Many of the decisions of adulthood rely on such complicated models of the social and economic world. Should I buy a house, or rent? How much of my salary should I put in my retirement account? How much should I give to charity? Should I try to write a book? You can build good models for some of these questions by thinking and reading, and by talking to people, but some things you just have to learn by action and experience: your particular situation will not be quite like anyone else's.

Lots of communication is about collective model-building: when we agree, we have the same model. Or, since there is very little that we can be absolutely sure of, we have at least convinced ourselves that our models agree. When we turn out to be incorrect, we eventually might discover that the candidate is not such a good fit for that job, or we don't have fun on a date.

When we are pleasantly surprised, sometimes this makes our lives more interesting and exciting: like when my first-grade daughter decided that her favorite song was the Beatles' "Yellow Submarine" (finally beating out "Let It Go" from Disney's *Frozen*).

This model-building gets formalized in scientific discourse, with papers or conference talks presenting the scientists' latest version of their models, explicated precisely enough for their colleagues to see and compare with their own. An experimental paper may be presenting data that back up, or contradict, the available models; a theoretical paper may be presenting a new model or drawing out the repercussions of an extant one.

In 1953, *Nature* published three articles that collectively set out the double helix structure of deoxyribonucleic acid, or DNA, the molecule that encodes the genes of all Earth's living creatures.[3] The paper by James Watson and Francis Crick recounts the discovery of DNA's structure, but in fact it is almost entirely based on the experimental work of the authors of the other two papers, in particular Rosalind Franklin, who had worked out how to best use the technique known as X-ray crystallography to determine the structure of complex biological molecules. Watson himself later said that "the Franklin photograph was the key event." That was the crucial information that allowed Watson and Crick to build their model, the double helix, which unlocked our understanding of genetics.

But this was not a great example of scientific communication. The photograph had been shown to Watson and Crick by Franklin's King's College London colleague—and competitor—Maurice Wilkins without her permission; Wilkins and Franklin were the respective first authors of the two other papers in the trilogy. Alec Stokes, one of Wilkins's coauthors, also at King's, had worked out some of the helical model a couple of years earlier but had left it unpublished until it was discussed in the 1953 paper with Wilkins. Models are judged not only by their accuracy but by the uses to which they are put.[4]

How Do I Know What I Know?

We know about the dinosaurs because we have pieced together their fossilized remains, analyzed their chemical composition and the locations in which they were found, and put together a *model* for how they lived and evolved.

I know a song I hear is by the Beatles because I recognize John's or Paul's or George's (rarely) or Ringo's (hardly ever, but easily recognized) voice, and their harmonies, and the kinds of songs they wrote, and the way they were produced. I have a *model* of the Beatles in my head. I know the schedule for my bus, which is just a *model* for how often it should arrive. I know when my baby daughter is hungry because I have a *model* of her (and babies in general) in my head, perhaps partially generated by millennia of evolution to ensure that we wouldn't let our offspring go hungry.

Some of these models came from our attempts to understand the data at hand: the light from the stars in the sky, fossilized bones, the sounds coming out of the speaker. And it is just in the interplay of the model and new data that we can learn something. We adjust the model to fit the data, and we collect more and more data to refine the model further.

Sometimes models are wrong, of course: we didn't understand the evolution of dinosaurs into birds until the late twentieth century; the song that I think is by the Beatles might be by the Kinks or the Rolling Stones; the buses on my route might be canceled. Sometimes, these errors might lead me to correct my model, but often I understand that the model is not exact. So it will lead me to make mistakes, but I want to keep those mistakes to a minimum. Even better, I want to understand something about those mistakes: How often will they occur? What kind of mistakes are most likely? I am more likely to mistake a Beatles song for one by another 1960s group than for one by Metallica or the Sex Pistols. I might mistake a small fragment of fossilized bone for a dinosaur from the Cretaceous period rather than the Jurassic, but probably not for a piece of fossilized wood. I might confuse a nearby star with a very distant galaxy, but not a comet.

So when I compare a model with data, I not only want to know the answer to some question (Who wrote the song? How old is the star? Is the baby hungry?) but also how confident I should be in the answer.

This is much the same process whether the model describes the nuclear reactions in the interior of the sun, or the group dynamics of the new people you might have met at a party. This interplay between model and data is really the only way that we can answer *any* factual question about the world.

All these models, from describing the Beatles to crying babies or the Big Bang, are very sophisticated, although it wouldn't always be easy to articulate the details of my model for the Beatles, or of my children's behavior, in a precise way. That's what makes scientific models—of the sun, stars, galaxies, and the Universe—special: they are built up from small pieces and can often be described in an explicit mathematical form. That makes it easy to apply them to and, most importantly, check them against the results of scientific observations.

These theories let me put together a model for how a star shines and how one of those stars, the sun, 93 million miles away, can be bright enough to warm Earth and provide energy for life. The model works because it agrees with those underlying physical theories—gravity, nuclear physics, light— along with all the data we have gathered about the sun and stars. The models let us interpret the data, and the data help us refine our models. And there is still much to learn about the details of our own sun's structure and evolution and about all the different kinds of stars in the Universe, each slightly different from all the others—but each obeying the same underlying physical laws.

So these must be pretty complex models, involving lots of the stuff that I had to learn while getting a PhD in astrophysics and in the years after. But what's present in these models is the same sort of reasoning that we need every day. I start with those facts on the ground—the time according to my watch, the sound of crying, a pinpoint of light from a particular direction in the sky.

To interpret these observations I need a model. This can be complicated and technical, such as an astronomer's understanding of how stars evolve and how nuclear reactions and gravity work. Or it can be my more prosaic knowledge of London buses and traffic. Or the simple fact that sounds coming from a wall usually mean that something is making noise on the other side, and that a crying baby (or, with the knowledge gained from living with the baby the first few months of its life, at least *my* crying baby) is often hungry. Or our awareness that smiles and laughter usually (but not always, especially in Britain) mean that the people around you are happy.

The observations and the model together let me piece together something new: an understanding of a particular situation, often alongside a better understanding of the model itself. The more often the baby is quieted by her bottle, the more I can tell the difference between her reactions to hunger and a soiled diaper.

The Scientific Method

The process of model-building might sound familiar. In school, we all learned about something called the scientific method. A scientist has a hypothesis, some idea about the world. She designs an experiment to test this hypothesis, and the hypothesis stands or falls based on the outcome of this experiment. If the hypothesis withstands enough of these tests, we might start to call it a theory, but this is a fraught term with sometimes contradictory meanings.

The early twentieth-century astronomer Arthur Eddington is supposed to have said that one should "never believe an experiment unless it's confirmed by theory": you can't make sense of an experimental result unless you have some model to interpret it, any more than you can make sense of the sensations from your eyes and ears without at least a vague model of the world to guide you.

Eddington probably never said this, or at least he does not appear to have written it down anywhere. What he did write, somewhat less pithily, was:

> Observation is not sufficient. We do not believe our eyes unless we are first convinced that what they appear to tell us is credible.
>
> It is better to admit frankly that theory has, and is entitled to have, an important share in determining belief.[5]

Eddington was mostly a theoretical astrophysicist, but he might be best known for an experiment showing the scientific method playing out in almost exactly this way. In 1919, Eddington was one of the few physicists who understood the implications of Einstein's general theory of relativity.

This famous theory is a model for gravity: it says that the presence of matter warps space and time and hence alters the paths of particles. In most situations that we encounter, general relativity makes predictions about the motions of objects that are almost identical to Isaac Newton's simpler seventeenth-century theory of gravity—that's why his ideas had been accepted for the previous two hundred years. But Einstein predicted at least one startlingly different phenomenon, now called *gravitational lensing*—light rays don't travel along straight lines. Instead, a beam of light passing by a massive object will be bent by an amount determined by the object's mass and its distance from the beam.

Newton's older theory was not completely clear on this matter. In the usual way that Newton's laws were written down, the force between two particles depends on both of the particles' masses, and so we might expect the path of light—supposed to have zero mass—to be unaffected by the presence of another massive object. But Newton also showed us that the motion of any particle is determined by the mass of the objects around it, but does not depend upon its own mass. So the laws do predict that the path of a moving particle, no matter its mass, is bent from a straight line by the gravitational pull of a nearby heavy object, for exactly the same reason that the force of gravity bends the trajectories of planets into orbits around the sun.

The formula for how much the path of a particle bends depends on how fast it is moving, but not on how heavy it is. So perhaps light, traveling at a fixed speed of about 186,000 miles per second—and despite being assumed to have no mass at all—would be deflected the same amount as anything else traveling at that same speed. This argument was first made in an unpublished "scrap" by British physicist Henry Cavendish in the 1780s and then more formally by the Bavarian astronomer Johann von Soldner in the early 1800s.[6]

(Newton himself might not have agreed with this interpretation of his laws: one of his other well-known works was a book called *Opticks*, in which he championed a theory of light as composed of waves, which might not have reacted to gravity in the same way as particles. We now know that light, as described by the theory of quantum mechanics, has both wave-like and particle-like properties.)

So, perhaps, Newton's theory of gravitation does predict some sort of lensing of light rays. But, crucially, the amount of deflection in this interpretation of Newton's theory is a factor of two smaller than Einstein's prediction. Figure 1 shows the geometry of gravitational lensing in either theory.

But light moves very quickly, and the faster an object moves, the less it is deflected by gravity, according to both Newton and Einstein. In both cases, along with the speed of light, the amount of deflection depends upon the distance to the deflector and its mass: the angle gets larger for heavier masses and closer approaches.

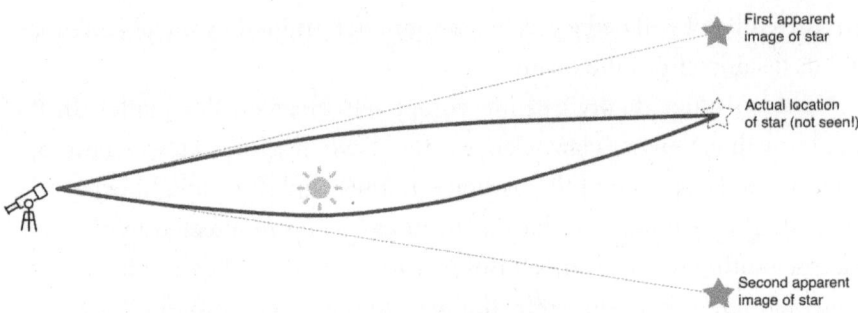

Figure 1. Gravitational lensing of a distant star by the sun. The star appears to have more than one image, at different points on the sky.

At first, then, this gravitational lensing was thought to be unobservable — no one had ever been able to observe Newtonian lensing of light by mass, because the amount of bending was so small in most circumstances. Even the mass of Earth itself is too small to make an observable change to the path of light. It would take something at least as massive as a star, and paths passing very near its surface, to divert the beam by an observable amount. A light beam grazing the surface of our sun would have its direction changed by just around one arc second (that is, 1/60 of an arc minute, or 1/3600 of a degree) — a little bit more for Einstein's prediction and a little bit less for Newton's. Unfortunately, a star is bright, so bright that it is usually impossible to see anything that close to its surface.

Although Einstein didn't complete his theory of general relativity until a series of papers in 1915–1916, he had already shown by 1913 that his theory would predict some deflection of light; at that point in the development of his theory, he had incorrectly calculated the same amount of deflection as the tentative Newtonian prediction. Astronomers had by then realized that a solar eclipse — the moon blocking the light of the sun — provided the ideal laboratory for testing these ideas. By comparing the relative position of stars during a normal nighttime observation to their positions during an eclipse, they could see the bending due to the presence of the eclipse-darkened sun. And a one–arc second deflection was just about observable given the instruments and techniques available at the time, by averaging over the noisy measurements of many stars. Unfortunately, the first expeditions to test these observations, by Germany, the US, and Venezuela, were scuppered by cloudy conditions—and the onset of the First World War. Scientific collaborations are difficult to conduct between hostile nations.

Eddington was a Quaker and a conscientious objector; he was granted a deferment from military service and allowed to plan an expedition to observe a postwar eclipse. Astronomers knew there would soon be an opportunity: another solar eclipse would occur in the middle of 1919. The British set to work, planning expeditions to Principe, off the west coast of Africa, and to Brazil, to measure the locations of the stars on the sky during the eclipse. After analyzing their data, Eddington was able to show that the calculation from Einstein's theory fit the data and Newton's did not. Both *The Times* in London and the *New York Times* (where it appeared as a "special cable") noted Einstein's seemingly unequivocal triumph (and Eddington's).

(There has occasionally been some doubt about the quality of Eddington's observations, suggesting that he might have fudged his data analysis in favor of Einstein's prediction. Modern reanalysis of the data has indicated that there was no foul play, or even any "confirmation bias"—the possibility that Eddington so wanted Einstein to be right that he inadvertently made subtle errors that nudged his result in what he thought to be the right direction.)[7]

Of course, the 1919 eclipse didn't *prove* Einstein's theory; all that observations during the eclipse could do was to make relativity seem much more plausible than the older Newtonian theory. In following the reporting style of all good scientific results, Eddington and his collaborators didn't just report a single number for the measured deflection; they also gave an estimate for the trustworthiness of their results, in the form of an error bar. They reported a measurement of 1.98 arc seconds, with an error of 0.18 arc seconds.[8] Einstein's prediction was 1.7 arc seconds; Newton's was 0.85 arc seconds. Einstein's theory was clearly favored by these results, just over 1.5 error bars from the result, whereas Newton's theory was more than six error bars away (and the other possibility, no deflection at all, was more than ten error bars away!). We should never expect a real measurement to exactly agree with a prediction, even from a correct model: noise will inevitably perturb the measurement from the truth. So Einstein's much closer prediction had won the day.

Years later, Eddington used the example of gravitational lensing to show how all our observations need at least some theory to interpret them: Einstein, and Eddington himself, showed us that mass deflects light. But how much? Once you accept the possibility of lensing, you then need to be sure that most astronomical observations are *not* lensed; in his description, Eddington starts by reminding us that we don't have any direct knowledge of stars other than the points of light we see in the night sky:

> For the reader resolved to eschew theory and admit only definite observational facts, all astronomical books are banned. There are no purely observational facts about the heavenly bodies. Astronomical measurements are, without exception, measurements of phenomena occurring in a terrestrial observatory or station; it is only by theory that they are translated into knowledge of a universe outside.
>
> When an observer reports that he has discovered a new star in a certain position, he is probably unaware that he is going beyond the simple facts of observation. But he does not intend his announcement to be taken as a description of certain phenomena that have occurred in his observatory; he means that he has located a celestial body in a definite direction in interstellar space. He looks on the location as an observational fact—on a surer footing therefore than theoretical inferences such as have been deduced from Einstein's theory. We must break it to him that his supposed "fact," far from being purely observational, is actually an inference based on Einstein's theory—unless, indeed, he has based it on some earlier theory which is even more divorced from observational facts. The observer has given a theoretical interpretation to his measurements by assuming for theoretical reasons that light travels through interstellar space approximately in a straight line.
>
> . . . So if we are willing to use both fact and theory as a basis for belief, we can accept the observer's announcement; but it is not a "hard fact of observation."[9]

Eddington is reminding us that, absent at least some theory to interpret the spots of light we see through a telescope, we would not really have any reason to believe that light travels in a straight line over vast distances and that only by dint of the relatively small concentrations of matter in the Universe, most of the light that we observe through our telescopes is *not* distorted by gravitational lensing. In a more complicated universe, we might have to contend with much stronger lensing, and perhaps never be sure of the locations of distant objects, as if observed through fun house mirrors.

In fact, we now know that a small but significant portion of our astronomical observations are affected by gravitational lensing and that this must be taken into account when interpreting them. Some objects are made brighter, some have their shapes distorted, some have their positions changed, and any of these effects can make us misinterpret what we are seeing: the galaxy may be farther away from us or from its neighbors than we thought, or it may be rounder than we thought. And these distances, and shapes, and neighbors are some of the tools we use to determine the properties of the Universe on the very largest scales, measurements that we will simply get wrong if we do not account for the distortions of gravitational lensing.

While I was writing this book, NASA launched the James Webb Space Telescope, the largest telescope in space. Its very first image, shown in Figure 2, was a high-resolution picture of a cluster of galaxies 4 billion light-years distant from Earth. The mass in the cluster of galaxies is so large that we can very easily see the distortions to the images induced by gravitational lensing. The shapes of the background galaxies are stretched out, tracing the mass of the foreground cluster.

Einstein revised Newton's theory, leaving enough intact that the earlier theory's predictions remain valid in most situations, but making new predictions that Eddington, and many others since, have been able to verify.

A century after Eddington's analysis, our current model says that Einstein's theory of gravitation is correct, but it does not predict the distribution of matter in the Universe—the source of both light and gravity. General relativity tells only how this distribution evolves over time, but not how it began. Without predictions for those initial conditions, we have to rely on observation and experiment: we need more information.

Figure 2. A cluster of galaxies (called SMACS 0723) as observed by JWST in 2022. The fuzzy, stretched-out galaxies are in the distant background, and their images are distorted by gravitational lensing by the cluster in the foreground. (NASA.)

By leaving our model open-ended in this way, we can self-consistently use our observations of gravitational lensing to detect the presence of matter and make quantitative statements regarding the locations and masses of objects such as galaxies and stars. Such open-ended models are the way science proceeds: we use them as the basis for designing experiments and interpreting their results, making measurements of those unknown quantities.

In Eddington's time, gravitational lensing was barely observable, so Einstein's theory confirmed that the Universe was more or less as we naively perceived it. Nowadays, we can use the predictions of Einstein's theory to help us detect and measure the distribution of the mass that is responsible for the lensing—one of the challenges and opportunities for astronomy and cosmology in the twenty-first century.

But we will return again and again to Eddington's broader point, that all our scientific observations—all our perceptions, many of what we usually call "facts"—are laden with theory and models.

What Is This Thing Called Science?

Worrying about the deflection of light rays when making maps of the stars seems esoteric, but this need for theory to interpret even the most basic observation goes back to the dawn of the scientific method. The ancient philosopher Aristotle had seen that all moving objects eventually slow to a stop. So he elevated this commonplace notion to a principle: the natural state of objects is at rest; only if a force is continuously applied will an object continue to move. This, along with many of the rest of Aristotle's philosophical ideas, held sway among the intelligentsia in Europe, the Middle East, and parts of Asia for nearly two thousand years. With the European Renaissance and Enlightenment, those ideas at first gained more prominence, but so did the notion that the ancients might not have been right about everything. So, in the seventeenth century, Galileo Galilei began to think about the movement of objects on Earth and in the heavens. And he realized that Aristotle was wrong, despite every single experiment ever performed having shown that objects do, eventually, slow down and come to rest. But through even more careful observation, Galileo realized that only because of the ubiquity of what we would now call friction (and related phenomena such as air resistance) does everything seem to slow down. As you decrease friction, moving from sliding along grass to a polished surface to a flat, icy pond, you increase the distance and time over which an object will keep moving. And so Galileo did an amazing thought experiment, one that would be impossible to put into practice: What happens if you continue to decrease the friction, all the way to zero? He realized that the object would move farther and farther, slowing down less and less. Instead of slowing down, an object subject to no force whatsoever—neither being pushed along nor being slowed down—would continue to move, forever. Galileo upended Aristotle, just as Einstein would do to Newton.

But as the twentieth-century philosopher of science Paul Feyerabend put it, "It was the Aristotelian theory of motion which was defended by reference to experimental results, while Galileo who was not prepared to take these results at their face value, but insisted that they be analyzed and be shown to be due to the interplay of various and as yet unknown factors."[10]

Galileo and Newton were among the first modern Europeans engaged in what we now call science (a word that didn't come into its current use until the late eighteenth century, when it supplanted the earlier term "natural philosophy").

If I had to define "science" in a single sentence, I would say that science is the process of making inferences about the world from data.

Let's see if we can pick this description apart. First, science is a *process*, a way of looking at and trying to understand the world. It's not a body of knowledge; it's not the answer to some questions; it's not just theories or hypotheses or experiments. It is an ongoing activity that uses those theories and hypotheses and experiments to answer questions and build up bodies of knowledge.

Taking the terms slightly out of order, the process of science relies on *data*. These data are often the result of an experiment, although it doesn't need to be as explicit and controlled as that.[11]

Finally, that process uses the data to make *inferences*, which the Oxford English Dictionary somewhat vaguely defines as "conclusions based on evidence and reasoning." The *Merriam-Webster Dictionary* splits the definition in two:

1. the act of passing from one proposition, statement, or judgment considered as true to another whose truth is believed to follow from that of the former;

2. the act of passing from statistical sample data to
   generalizations (as of the value of population parameters)
   usually with calculated degrees of certainty.

Inference, considering all these definitions, has repercussions for the way science is done. First, we are starting with things considered as true. That's an important hedge: not "true" but "*considered as* true." In our scientific method, we don't really have a way to get to absolute, incontrovertible truth. Instead, we choose a starting point and see where the inference leads. Of course we can try to reduce the list of those things we are *considering true* to be as short as possible. Even better, we can use the method to help *assess* their truth, by searching for contradictions: predictions that are not borne out by further experiments or observations.

What sorts of statements should we take as these "propositions considered as true"? First, we must assume the truth of our data. This doesn't mean that our measurements must be perfect, but just that we have a good description of how they were taken and how they might have gone wrong. We are used to seeing scientific measurements with error bars, and this is where they come into the process. With those error bars, we can associate a number with how far we think we might be from the true quantity. We might trust the speedometer on our car to be accurate within one or two miles per hour, and we might trust a modern watch to much better than a second.

Second, when we are interpreting those data, we tentatively assume the truth of the *models* that I have claimed underlie all scientific reasoning and much of human learning and behavior besides. We take these models to be true and try to refine them or, if they seem to be in contradiction with the data or our other assumptions, reject them. So the models may not, in fact, be true, but we start by making our inferences *as if* they are.

One very important part of the practice of science is making approximations: simplifying our models so that we can understand and apply them, even if that means we know they are, technically, false. We might treat Earth as a perfect sphere even though it bulges out slightly along the equator, not to mention the variations in its shape due to oceans and mountains. We might try to model only the results of an election, rather than the preferences and votes of every individual citizen. We might know that our model only captures enough to predict the weather over the next few days but no longer. We might assume that galaxies, despite comprising tens of billions of stars, may as well be infinitesimally small point masses when we assess how they behave in the vastness of the expanding Universe. If these approximations hold well enough, our model might still be useful.

That our models are "useful" in this way may seem like not much of a goal, but that is exactly the point: a model can be useful if it gives us an approximate description of what we are studying—geography, climate, politics, or astrophysics—especially if it also gives us some information about how the approximation can fail.

We don't need the exact time the rain will start to know that we need to carry an umbrella. We don't need to know the precise trajectory of a ball to be able to catch it, but we can increase our chances of catching it if we can account for gusts of wind.

The second definition of inference from the *Merriam-Webster Dictionary* is more specific to its use in science: inference "generalizes" from data with "calculated degrees of uncertainty." This is just a roundabout way of saying that these generalizations should come attached with a *probability*, a number calculated to describe how certain we are in the truth of the generalization. This is how we cope with these "wrong" models—we can never get to 100 percent certainty, so we ask that our method gives us a way to generate these probabilities, rather than precise answers.

Okay: I need a model to interpret the data from my senses. But where do I get these models to begin with? I might have learned some of them from my parents as a child or in school, and some might even be innate, built into my physiology by billions of years of evolution. Many seem to come out of that data itself. Humans have observed the sun come up every morning, that it's cold and dark in winter, and warm and sunny during the summer, and that this pattern repeats every 365 sunrises and sunsets, and somehow we have created a model that these patterns will continue, even without a concrete physical model of the planet and the solar system.

When we are born, we don't understand the way the world works, even in very basic ways. But after around eight months of experience, as babies we figure out that the world is composed of discrete things, objects that can move around but don't randomly appear and disappear. This is the onset of "object permanence," one of the early milestones of human development as described by psychologist Jean Piaget.[12] We have somehow built up a simple physical model of the world based on the experience of our eyes and other senses, and of our experiments—our interventions into the world, pulling and pushing and playing.

Science uses inference to refine our models, plan our experiments, and interpret our observations. Like an infant's tentative attempts to learn about the complicated external world, the scientific coupling of models and observations is a living, breathing, subjective endeavor, simultaneously collective and personal.

TWO

The Problem and the Promise of Induction

All the models that I've described, from crying babies to gravitational lensing, are built in the same way. We have observed some sort of regularity in the world, and from that we have extrapolated that this behavior is widespread, or even universal, a process of generalizing from finite amounts of data.

Another word for this process is *induction*. Induction is defined as "the inference of a general law from particular instances." Without induction, or at least without those general laws that induction produces, the world would just be a cavalcade of uselessly unlinked facts. Those general laws allow us to extrapolate from what we see, to places we haven't been, and to the future. With these extrapolations, we can act: wake with the sunrise, plan for winter, or just assume that the ball rolling across the floor will make it over for us to play with.

Induction helps us figure out what to believe and guides our actions in light of those beliefs.

Induction vs. Deduction

The Scottish Enlightenment philosopher David Hume asked how we can know about causation, how one thing makes another thing happen. The answer, he wrote in both *A Treatise of Human Nature* in 1739 and *An Enquiry Concerning Human Understanding* in 1748, is that "we can go beyond the evidence of our memory and senses." Those senses tell us, literally, about the things in front of our noses, which directly affect us. Everything else, Hume thinks, is supposition based on a "chain of reasoning."

Hume preferred *deduction*, the process of logical proof, in which we start with axioms and tease out all their repercussions. The word "axiom" comes from the Greek, and many will have first seen it in studies of Euclid's geometry. These axioms are a sparse set of what are essentially definitions of lines, points, circles, and angles, and from there we can deduce (prove) that the shape of a triangle is defined by two sides and the angle between them, or even the famous Pythagorean theorem, that the square of the hypotenuse of a right-angle triangle is the sum of the squares of the lengths of the two other sides. Since these ancient Greek philosophers and mathematicians, the notion of a "proof" was believed to supply thinkers in the West with tools to determine if something was true or false. (A "theorem" is the name we give to the more interesting statements that we prove from those axioms. This is etymologically related to, but distinct from, "theory," a word that we use to describe the most interesting of our scientific models.)

Induction, by contrast, is the mental process of making general statements from more specific ones: the sun has risen every day of my life (and for many more days before I was born) so, supposedly based on just those facts, induction implies that it will do so again tomorrow.

The seventeenth-century French philosopher René Descartes saw the problem with this sort of reasoning: How can I be sure of anything at all, even my own senses? Hume crystallized this fear into a problem of logic: Can I *prove* that the process of generalizing from my (supposed) sense impressions is sound?

Of course not—Descartes was right. The sounds outside my office, *I reason*, are evidence of someone on the other side of the door, because those are the kinds of sounds that people make when they are on the far side of a door. But how do I *know* that those particular sounds are produced by a person? Because all the other times that I've heard such sounds, they were produced by a person. But what about this particular time? How can I be sure? I seem to need to invoke a new principle, "that the future will be conformable to the past" in Hume's words. How can I justify this?

Hume himself calls this thought process "probable" reasoning, already showing us the way to our goal. However, lacking the tools to describe probabilities mathematically, Hume in fact saw this not as a solution to his dilemma, but part of the problem. Probability will never give us certainty, and the desired language of proof is the classical language of reason and certainty.

Starting from the truths that Socrates was a human and that all humans are mortal, an ancient philosopher could use *deduction* to prove that Socrates was mortal, just as Euclid could prove statements about shapes and numbers. So this *deductive* process of proof seemed like a way—perhaps the only way—to generate new truths, statements we can be certain of. Induction is more subtle: Does the fact that "all the humans *I know about* are mortal" imply the broader statement that "all humans are mortal"? No—to prove the second statement, you need to know biology at least.

Deduction is logically sound, but this fact makes it a little empty. It doesn't feel like we are learning anything truly new when we apply it. Induction is more exciting—because we seem to be creating new truths when we use it. But this means that it can fail: we might have just missed a counterexample to our generalization.

Hume saw this as a problem. Many of the things that we understand to be true don't follow from incontrovertible statements. If I hold something heavy and drop it, it will fall. I know this because every time I've dropped an object in the past, it has fallen to the ground. I would like induction to let me prove that all heavy objects fall. But this is exactly backward from what I can do with deduction. If I already knew that all heavy objects fall when dropped, and I knew that this rock I held was a heavy object, then I can deduce that the rock would fall if dropped. Nowadays, we have a physical model—gravity, via Newton or Einstein—we can use to understand why objects fall to the ground, but absent that, there seems to be no purely logical justification for believing that heavy objects will fall to the ground.

The Socratic example already shows us the problem. Where does the statement "all humans are mortal" come from? It's certainly true that I don't know of any humans more than about 120 years old, so, to the best of my knowledge, most humans who have ever lived have also died, and so are certainly mortal. But we don't know, for sure, that everyone today is going to die, someday. (There are futurists and billionaires who want to think otherwise, in a long, so far unsuccessful, tradition stretching back at least to medieval alchemists and Spanish conquistadores searching for the elixir of life.)

We could take mortality as a given fact, part of the definition of what it means to be a human. But this strategy would seem to lock us out from learning a lot of new and potentially interesting things about the world, unless we were somehow given those truths a priori, which would kind of defeat the whole purpose. Not all the truths that we can deduce are completely self-evident: Euclid's proofs certainly tell us things about geometry that are not obvious from the five postulates. But deduction still sharply limits the kinds of things we can learn.

So it seems impossible to prove that induction is correct. Perhaps worse, it looks like I might need to use induction exactly to get the kinds of axiomatic statements that I need to apply the supposedly ironclad logic of deduction to the real world, which doesn't come to us complete with a set of postulates or axioms from which we can start.

Hume saw that, therefore, we can't be really, truly, absolutely sure of any such generalizations—a counterexample may always be lurking for us to find.

We might try leveling up and noticing that the process of induction itself has always worked in the past, so perhaps that is some sort of justification for it. Philosophical opinion is divided about whether this is just a question-begging vicious circle or whether the fact that it's a sort of virtuous spiral of a metastatement buys you anything.

This is *the problem of induction,* so named by John Stuart Mill in 1879: Does reason alone allow me to generalize from some small, or at least finite, set of observed facts to more general statements about the world?

The answer, Hume worried, is no: we simply cannot reason our way to general statements from specific ones. This must be, strictly speaking, correct. We cannot always generalize from what we have seen to what we have not. Not all swans are white, and no number of past sunrises will make tomorrow's absolutely certain. (Someday, in the very distant future, the sun's nuclear fires will be extinguished, and with that, most likely, Earth will perish.)

We can extend this skepticism even more radically: Why should I even believe the facts that appear to be right in front of my face? Might I be deceived even about them? Why should I "generalize" from the relatively small number of things that I have seen with my own eyes to the existence of anything else at all beyond them? Perhaps I'm dreaming, or under the influence of hallucinogenic drugs? Could I be a "brain in a vat" being electrically stimulated to make me think I'm living in a world that is entirely illusory?[1] Could we be living in a simulation as in the *Matrix* films, or be characters in a vast computer game? None of these are logically impossible, and this leads to the kind of radical skepticism articulated by Descartes, who decided that the only things he could be completely sure of were his own thoughts: *Cogito, ergo sum,* he wrote in Latin (I think, therefore I am).

So we seem to be in a pretty dangerous place. A scientific theory is exactly the sort of generalization Hume worried about justifying.

In cosmology, we have an extreme version of this problem due to the finite speed of light, putting it in the most elegant way possible, at one light-year per year. A light-year is a measure of distance—about 9.5 trillion kilometers or 4 trillion miles. We observe the distant Universe entirely using light reaching us from distant stars and even more distant galaxies.[2] But that means that we are only directly aware of what's happening on a series of nested spheres: things that happened one year ago on a sphere with a radius of one light-year, or ten years ago on a sphere of radius 10 light-years, or a million years ago on a sphere of radius 1 million light-years, and so on, back to the time of the Big Bang. This nesting of spheres is called our light cone. We can get signals via slower-moving objects or particles that can fill in those spheres, but the absolute universal limit of the speed of light means that we can't have any information at all about stuff happening outside the spheres (outside the light cone). The Universe is a pretty big place, and this means that we're missing direct information about almost all of it. Not because we aren't smart enough, or because we haven't built good enough telescopes, but because that information is fundamentally out of our reach due to the finite speed of light. The rest is only accessible to us via induction, via our theories.

So Hume's suggestion shows us the solution to the problem of induction: "the future is conformable to the past" is a model, exactly the kind of mental tool that we need to learn about the world. But it's not quite precise enough to be useful. We need models that are explicitly tuned to the questions we are asking.

The whole enterprise is a kind of category error, trying to show that induction can somehow be converted into, or reduced to, deduction. The opposite is true: deduction is a special case of induction, applicable when we are dealing with things we already are certain to be true. Induction lets us deal with uncertainty, if we are willing to account for the probabilities that Hume was worried about.

Naturalism

One strange thing about "the problem of induction" is that no one actually thinks it's a problem. Well, at least no one learns about it in philosophy class, throws up their hands, and decides never to use induction again. That would be impossible. Induction is not only how we learn that Einstein's theory of gravitation seems to be correct. Induction is part of why we eat when we are hungry, why we close our eyes when it is too bright, why we wear clothing to protect us from the elements, why we come home expecting our family's love and a hot meal. As Frank Ramsey, an amazing figure who died at 27 in 1930 after making remarkable contributions to philosophy, probability, and other branches of mathematics, put it,

> Induction is such a useful habit, and so to adopt it is reasonable.
> All that philosophy can do is to analyse it, determine the degree of
> its utility, and find on what characteristics of nature this depends.[3]

Induction makes sense if—and only if—the world makes sense. This is the idea of "naturalism"—the overarching model that the world is, at bottom, intelligible. Naturalism is more or less built into the fabric of our beings by billions of years of evolution; the life of a bacterium is no less dependent on the regularity of the world around it than we are, even though it is little more than a biological machine, lacking in self-awareness. The bacterium doesn't know that it is using induction, but the evolutionary fitness of an organism depends upon its ability to make predictions about its environment. It needs to move away from danger and toward food and sources of energy. Induction is not just a way of thinking that we can turn on or off when we are doing science; induction is built into the way we interact with the world around us.

In 1949, Albert Einstein wrote that "The justification of such constructs [as causality do] not lie in their derivation from what is given by the senses. Such a type of derivation (in the sense of logical deducibility) is nowhere to be had. . . . The justification of the constructs, which represent 'reality' for us, lies alone in their quality of making intelligible what is sensorily given."[4]

That is, we begin with causality, and even more basic concepts like the existence of an external world. As Einstein writes elsewhere in the same essay, they are "indispensable elements of thinking." Put another way, one solution to Hume's problem is to understand *psychology*: we couldn't even think without induction.

We can take this realization further. If the solution is psychological, then it must also be a product of evolution. Our minds evolved to look at the world *as if* it were governed by causality. And this strategy works because causality, and the use of generalizations to make broad statements about the world, has contributed to our evolutionary success—causality seems to be true and induction works. (Of course, it remains *logically* possible that there really is no external world, that we live in a *Matrix*-like simulation, or that your brain has just mysteriously popped into existence in the last few minutes with its—false!—memories of the world somehow implanted.)

Two decades after Einstein's writings, the philosopher Willard V. O. Quine took this idea further in an essay called "Epistemology Naturalized" where he made the link between naturalism and psychology explicit—he suggested that the philosophical project of epistemology should be subsumed under the natural sciences of psychology and cognitive science.[5]

All this is true, but it doesn't really give us scientists enough to work with. After all, epistemology is the study of how we know what we know, and the goal of science is to increase what we know. Retraining as a psychologist would not help me become a better physicist and learn about the history and evolution of the Universe.

You could, I suppose, deny that naturalism *always* applies: there could be specific circumstances under which the universe is ungoverned by any natural law, completely without any sort of regularity that a living creature could take advantage of to further its needs. Perhaps the *super*natural is real and magic works. But a truly *un*natural kind of magic wouldn't even follow any sort of internal logic of its own. A supernatural being that only intervenes in the world utterly capriciously might fall into this category, but this being would not make much of a religion.

A formal philosophical definition of "naturalism" would probably distinguish between the scientific ("natural") and the "supernatural." But more broadly, if you believe that you can game the system to your benefit, then you believe that the Universe is governed by laws and at least some weaker version of naturalism applies. This is why astrologers try (and fail) to justify their divinations with (bad) science and why there has been so much effort to recast religious texts and beliefs as scientifically accurate. This is why the "magic" that appears in so many comic books and fantasy novels ends up seeming like a bizarre form of science, activated by spells and potions rather than fundamental forces and theories, but obeying strict rules nonetheless.

Of course, in the real Universe, we have found that even such magic does not, in fact, exist—and we've used induction to figure that out. So our real goal is not to justify induction, or to understand the cognitive and evolutionary processes that underlie it. We want instead to quantify induction, to figure out what mathematical rules we can use to describe how we generalize from our limited observations and how confident we should be.

THREE

What We Talk About When
We Talk About Probability

David Hume taught us that we can't make ironclad predictions. We can't logically, irrefutably, prove that we know anything about the world around us. But he misdiagnosed the problem of induction. We are bound to be confronted with uncertainty as we try to use our observations of that world to understand anything about it, and we can develop tools to use and quantify this uncertainty. We must use that "probable reasoning" that so frightened him.

This probable reasoning is probability, which allows us to work with statements when we don't know if they are true or false. In turn, the mathematics of probability will show us how to build models from smaller blocks, learning more about the world with each additional piece.

We all have some experience with probabilities. The weather app on our phone tells us there is a 30 percent probability of rain this Thursday. A pollster tells us that Hillary Clinton had a 90 percent chance of winning the US presidential election in 2016. Insurers and demographers interpret actuarial tables to tell us that a male in the UK has a 40 percent chance of living until the age of 88.

These numbers—30 percent, 90 percent, 40 percent—quantify the chance that something will happen. We can also interpret these numbers in a more general way: the statement "it will rain Thursday in London" has a 30 percent chance of being true; given what we knew in October 2016, the statement "Hillary Clinton will win the US presidential election" had a 90 percent chance of being true; and the statement "a male living in the UK will live to the age of 88" has a 40 percent chance of being true.

Instead of attaching probabilities just to the occurrence of events (the weather, the outcome of an election, death), we can now attach them to any statements that could be true or false, assigning numbers based on our knowledge. If I am sure that the moon is between 200,000 and 300,000 miles from Earth, I would assign a probability of 100 percent to that statement. As I write this, I haven't looked up the exact distance, so I am only about 50 percent certain of the statement "the moon is between 250,000 and 260,000 miles from Earth." (It turns out the moon is about 239,000 miles from Earth, so I was correct to assign a relatively low probability to the accuracy of the statement!)

Assigning numbers based on our knowledge lets us use probabilities in a more properly scientific context, where we want to make precise and quantitative statements, and for which we do not already know the answer. We might want to work out the probability that there is a planet around a distant star, say, or that a given drug is an effective treatment for a particular disease, or that the lemur shares a recent common ancestor with the loris.

To assign such a probability, we had first of all better define the terms of the statement very precisely. What do we mean by an "effective treatment"? What is a "recent common ancestor"? What qualifies as "a planet around a distant star"?[1] These definitions are part of the model that we are trying to understand better.

Once we've understood our model and the meanings of the terms of our statement, we will usually need some data—the results of a scientific experiment—and a way of checking the data against the model. For that, we will need to understand how to manipulate the numbers that we are using for our probabilities.

The Rules of Probability

For us to be able to use it in a coherent way, a probability must obey some simple rules. Let's start with a pair that are just definitions.

- "Probability one" (or 100 percent) means that something is certain to happen, or that is certainly *true*.
- Conversely, "probability zero" (zero percent) represents certainty that something is absolutely, for sure, *false*—it will never happen.

Because these are numbers, some readers may find it easier to use a more mathematical notation: $P = 0$ for false and $P = 1$ for true. We can think of this P as a sort of mathematical function, which takes some statement about the world and spits out a probability. For example I can confidently write

$$P(\text{the sun will rise tomorrow}) = 1$$

since I am sure this will happen. I can also confidently write

$$P(\text{pigs fly}) = 0$$

since we haven't yet genetically engineered winged super-pigs. (We'll talk later about the question of whether and in what circumstances I should be absolutely sure of even these things.) We want to be able to use probability to talk about any statement at all, so in the coming pages I will often not use the notation $P(...)$ for anything specific in the parentheses, but instead I'll pick some letter to represent *any possible statement*—just like we use variables such as x or y in algebra to stand in for some unknown quantity and write some function like $f(x)$—but rather I'll use a letter to stand for one of these statements. So, for example, $P(D)$ or $P(H)$ would stand for the probabilities of statements D and H. D could be the results of a political poll, for example, or H could be "pigs fly" in the example above, or something more properly scientific like "the Higgs boson has a mass between 120 and 130 GeV/c^2," in the strange units that particle physicists use to measure these things, or "the Big Bang happened between 13.2 and 14.2 billion years ago." I will try to switch back and forth between sentences and equations as things develop so we can get used to both kinds of notation.

Once we've pinned down these definitions of truth and falsehood represented by the values of probability one and probability zero, we can pretty easily work out some of the other rules that these probability numbers should obey. To do this, we need to add another sort of obvious ingredient, or a third rule, to the way we understand probability:

- The higher the number, the more likely it is to be true.

This fact implies an important corollary. Since the values of zero and one imply certainty, the third rule then implies that all probabilities are between zero and one, including the possibilities of exactly zero and exactly one.

All the interesting and useful things you can do with probabilities come from being able to find rules for combining the probabilities of different statements. The next rule says that, since a statement must be either true or false— even if you don't know which it is—when you add the probability of its being true to the probability of its being false, you get the number 1. That is

- The probability of the combination "the statement is true *or* the statement is false" is definitely true.

For example, the statement "Keith Moon played bass for the Kinks" is either true or false, so it is *true* that Keith Moon played bass for the Kinks *or* Keith Moon did not play bass for the Kinks. This statement has probability $P = 1$. Of course we have to define our terms carefully enough that the statement can only be either true or false; we would have to decide, for example, if Keith (who of course was the drummer for the Who) happened to be in the same recording studio as the Kinks in 1969 and randomly picked up a bass, and if he had, whether that counts as "played bass." This sort of imprecision of definition is different from the imprecision of probability: probability encodes the lack of knowledge that remains even when we know all the definitions.

We encode this "or" part of the statement about Keith Moon with *addition* in an equation. So, back to our mathematical symbols:

$$P(A \text{ or not } A) = P(A) + P(\text{not } A) = 1.$$

So, once I've chosen the probability that a statement (the A above, whatever it stands for) is *true,* I automatically know the probability that it is false: just subtract the probability of truth from one. So if the meteorologist says that there is a 30 percent chance of rain tomorrow, this corresponds to

P(rain tomorrow) = 0.30. By our rule above, this also means that P(no rain tomorrow) = $1 - 0.30 = 0.70$, or 70 percent. We can go further than this. If we have an exhaustive list of possibilities, we know that their probabilities must add up to one. We can split our "no rain" forecast into a few mutually exclusive possibilities: it could snow, for example, or it could be cloudy without precipitation, or it could be sunny. If those are the only possibilities, the probabilities for all these must add up to one:

$$P(\text{rain}) + P(\text{snow}) + P(\text{clouds}) + P(\text{sunny}) = 1.$$

For example, on a bleak February day in London we might have probabilities of 30 percent, 15 percent, 50 percent, and 5 percent for rain, snow, clouds, and sun, respectively, corresponding to

$$0.30 + 0.15 + 0.50 + 0.05 = 1.$$

It is up to you how to use this information. If you really don't like to get wet, maybe you bring your umbrella even when the chance of rain is 30 percent. Or if you really don't like to carry extra stuff, maybe you leave it at home even when rain is nearly certain. (This is the realm of *decision theory*, how to combine probabilities with our preferences to figure out what to do.)

We can also use this to help figure out what the probabilities might be, if we can add a little more information. If we have a standard cubic die with six sides, we know that

$$P(1) + P(2) + P(3) + P(4) + P(5) + P(6) = 1$$

where I've written $P(1)$ to mean P(the die roll comes up with one dot showing), $P(2)$ to mean P(the die roll comes up with two dots), and so on. In most cases, we know something else important: there is no reason to expect that any face of the die is more likely to come up than any other, so we should also have the information in the expression

$$P(1) = P(2) = P(3) = P(4) = P(5) = P(6).$$

Together, these equations tell us that each of those probabilities must be 1/6, since the six possible outcomes, each equally probable, must add up to one.

This way of assigning probabilities is often called "the principle of indifference," which has a long and at least occasionally fruitful history in probability, implicitly used by gamblers (and others) for hundreds (maybe thousands) of years. (Here, "indifference" does not mean that I don't *prefer* one outcome to another—I might very much prefer a particular roll of the die if I am gambling!—but rather that I don't have any reason to think that one outcome is more *likely* than another.) It was first articulated explicitly by the mathematician and physicist Pierre-Simon Laplace in the early nineteenth century and given its name by the economist John Maynard Keynes in the early twentieth. That is, in the absence of any other information, our probabilities should encode the information that we are indifferent to the possible outcomes—they should be equal. So, if there are two possibilities, they should each have probability 1/2; for three possibilities they should be 1/3. For a situation with some number N distinct possibilities, they should each have probability 1/N.

Of course, this principle is so intuitive that we use it all the time without bothering to give it a name. If something is equally likely to be true or false, we might expect that it should be given a probability exactly between $P = 0$ and $P = 1$: that just gives us $P = 1/2$. For example, in most cases the probability of a coin coming up heads or tails would be equally likely. It is why the probability of a coin coming up heads is 50 percent and why the probability is the same for tails, and indeed it's behind the strategies for almost all games of chance.

Finally, we also need a rule for the "and" part of a statement to complement the one for "or." Since "or" is annotated in the expression as addition, perhaps it won't be surprising that "and" (which I will sometimes shorten to "&") gets translated as *multiplication*. So, the probability that two statements are both true is given by their product:

$$P(A \ \& \ B) = P(A) \times P(B).$$

Just as we sometimes write multiplication by putting the two quantities right next to one another ($a \times b = ab$), we will also sometimes use that notation to represent the "and" part of the statement, and so we might also write another version of this rule:

$$P(AB) = P(A)\, P(B).$$

At this point in our discussion, this rule applies only to *independent* statements A and B where the truth of one doesn't depend on the truth of another. So, for example, if the probability of rain tomorrow is still 30 percent, and the probability that I will forget my keys when I leave the house tomorrow morning is 10 percent, then the probability that it will rain *and* I will forget my keys is their product, 0.30×0.10, which is 0.03 or 3 percent. If these statements were translated into expressions of probability, they would read

$$P(\text{rain \& forgotten keys}) = P(\text{rain}) \times P(\text{forgotten keys})$$
$$= 0.30 \times 0.10 = 0.03.$$

We can combine our addition rule for "or" statements with this multiplication rule for "and" statements to apply probability to more complicated situations. Let's consider rolling two dice. I want to know the probability of rolling snake eyes, both dice coming up with a 1 on their faces. Since the two events are independent of one another (the outcome of one die doesn't affect the other), I can just use the multiplication law:

$$P(\text{first die 1 \& second die 1})$$
$$= P(\text{first die 1}) \times P(\text{second die 1}) = P(1) \times P(1).$$

In the final statement, I've used the fact that we don't care about the order of the rolls of the die to equate $P(\text{first die 1})$ and $P(\text{second die 1})$ to the more general $P(1)$. Since each of those individual probabilities is 1/6, the combined probability must be $(1/6) \times (1/6) = 1/36$ —in other words there is a 1/36 possibility of rolling snake eyes.

But what if I want to know the probability of rolling a 2 and a 3? If I just wanted to know the probability of the first die coming up with a 2 and the second die coming up with a 3, this would just be $P(2) \times P(3) = 1/36$ again. But that's not what I really want to know; I am instead interested in the probability of either of the dice coming up with a 2 and the other coming up with 3. So now we have to combine the probabilities to work out the value of $P(\text{first die 2 \& second die 3, or first die 3 \& second die 2})$. To figure this out we can now use the "or" (addition) law: the probably of rolling a 2 or a 3 on either dice is 1/36, as noted above, but since that probability is for *each* die, now we add these probabilities together:

P(first die 2 & second die 3, or first die 3 & second die 2)
$= P$(first die 2) $\times P$(second die 3) $+ P$(first die 3) $\times P$(second die 2)
$= P(2) \times P(3) + P(3) \times P(2) = 1/36 + 1/36 = 1/18.$

This expression translates to a probability of 1/18 in rolling either a 2 or a 3 on either die, twice as likely as rolling snake eyes. The more general multiplication law is a little more complicated than this and will require us to expand the way we think about probabilities.

So far, I've managed to avoid pinning down one of the most important questions remaining about probabilities: Where do they come from? I talked about the simplest case, when I am sure that there is no reason for any of the outcomes to be more likely than any other—in that case, Laplace's principle of indifference applies, and we just assign all the possible outcomes the same probability.

In this case you might answer that it's just obvious: a coin has two sides, I don't know which one it will land on, so of course the probability that it lands on either side must be 1/2. It almost seems like this is a property of the coin itself, just as it's a property of a pair of dice that snake eyes will show up with probability 1/36, or, less precisely and in a different situation, that a straight flush is the least likely poker hand.

But let's think about the simplest of these examples, the coin flip, a little more. Consider the following cases:

- I've flipped a trick coin, either two-headed or two-tailed, but you don't know which.
- I've flipped a fair coin, and *I* have looked at the side facing up, but I have not told *you* the outcome.

In both cases, there is a right answer to whether the coin shows heads or tails—but you would still be smart to assign a probability of 1/2 to both heads and tails, since you are not privy to that information. On the other hand, if you peek through my fingers, you may be able to learn a bit more.

So the principle of indifference does not always make sense. It applies when there is some sort of symmetry between the different possibilities. The name of this principle is the key: when I think I should be "indifferent to" which of the outcomes is more likely, we can use it.

On an episode of the comedy/news television program *The Daily Show*, the reporter John Oliver discussed the possibility that the Large Hadron Collider at CERN might inadvertently create a black hole that could destroy the world. One of the interlocutors, not too well-versed in the underlying physics, but worried enough to have filed a lawsuit attempting to stop the collider from being turned on, was asked what the probability of such a terrifying outcome might be. "Well," he offered, "if you have something that can happen and something that won't necessarily happen, it's going to either happen or it's not going to happen, and . . . so the best guess is one in two."[2]

Even John Oliver responded, "I'm not sure that's how probability works." This example is the principle of indifference at work, completely inappropriately. In this case, we do have other information—the laws of physics, combined with the details of CERN's experimental setup—that tell us that the two outcomes are not equally likely. *The probability you assign depends on the model.*

We don't know everything about the laws of physics that are at work in particle accelerators—if we did, we wouldn't need to do the experiments to begin with! But we do know a lot, and all the calculations implied that the interactions at CERN could produce new and interesting particles, but not planet-eating black holes. Beyond theory, we have a lot of experimental evidence: every day, Earth is bombarded by cosmic rays, incredibly fast-moving particles from outer space.

The strongest cosmic rays are in fact more energetic than the particles we can produce at CERN or anywhere else. And those particles hit Earth every day and interact with its atoms in much the same way as inside CERN's detectors. And yet Earth is still here, not having been converted into a black hole. Even more, all the stars in the Universe are also bombarded by such energetic cosmic rays, and we have never once seen evidence of such a catastrophic interaction. So we know that the chances of it happening at CERN are very, very low. Perhaps not impossible—very few things are truly impossible—but so improbable that it would not happen in many times the age of the Universe.

Radical Subjectivity

Probabilities aren't features of the external world. Rather, they are *assigned* by those of us who use them, based on whatever information we have at hand, and wish to consider—our model. This was put rather starkly by the Italian statistician Bruno de Finetti: "Probability does not exist."[3]

He was contrasting the notion of probability to other scientific terms, which, we hope, correspond to things in the world that we can measure and agree on. Molecules exist, organisms exist, stars exist, independently of us. But probability, in de Finetti's view, exists only in the mind of an individual user. This is a scary sort of radical subjectivity that might make scientists uncomfortable. But I'd like to put forward a version that this subjectivity still retains a sort of objectivity, the only sort of objectivity that matters.

This requires us to expand our notion of probability and, yes, to introduce a little more mathematical notation. The new idea is called "conditional probability" and means you can't assign a probability without having at least some other information. For example, what information do I need in a coin toss to assign that same probability 1/2 to both heads and tails occurring? I need to believe that it's a standard coin with a head on one side and tail on the other (for instance, it's not a two-headed trick coin). I need to assume that the person tossing the coin isn't trying to cheat and flip it in such a way that he or she can control the outcome. I need to assume that we can neglect the possibility that the coin will land on its edge. Given (or, using the statisticians' term of art, "conditioned upon") those pieces of background information about the coin and its flipping process, it makes sense for me to assign probability 1/2 to each head and tail.

In symbols, to annotate this new condition, we now write

$$P(\text{heads} \mid \text{fair coin \& fair flip \& does not land on edge}) = 1/2.$$

The vertical bar in the equation ("|") is what marks this expression as a *conditional probability*, often read as "given"—in words, we read the expression as "the probability of heads *given* that it's a fair coin and tossed with a fair flip, and ignoring the possibility of it landing on its edge." That is, I need a model for the coin and the coin-flipping process before I can assign a probability for the outcome. Conditional probabilities are the link between models and uncertainty.

So I would propose a slogan that sounds somewhat less radical than de Finetti's: *all probabilities are conditional*.[4] Absent a model, I can't assign a probability, and absent a probability, I can't quantify my uncertainty about the world.

Until now I've been using a misleading notation: there's really no such thing as P(heads) *without* the conditional vertical bar used in the previous expression. I need a model for the coin and the coin flip before I can assign the probability. But if you and I agree on the model, we should assign the same probability. That is where we can strive for objectivity: if we are given the same information, we come to the same conclusions.

In our more generic formulation with mathematical symbols, a conditional probability could be written as $P(A|I)$, which means the probability of statement A given background information (or model) I. The letter A is just a placeholder for the statement we're interested in, such as "the coin will land on heads," and the letter I is a placeholder for any other statement or statements that we need to know in order to pin down the assignment of probability.

Imagine you are about to undergo a complicated surgical procedure, and you are told by your doctor that the surgery has a 50 percent survival rate but you should not be concerned since her last 20 patients all survived. Should you worry?

This scenario is taken from an internet meme, circa 2022, contrasting the response of a mathematician and a non-mathematician. A non-mathematician is supposed to think that their surgeon's survival rate tells them to ignore the overall low survival rate and be confident in their surgeon's skills. But a mathematician is supposed to realize that the 50 percent survival rate means that their chances of survival are still fifty-fifty, despite their surgeon's record.

In fact neither response is right. At this point in the chapter, you probably won't be surprised to hear that the answer to the question of whether you should worry depends on—is conditional upon—your model. The fifty-fifty response from the mathematician makes sense only if you believe that the only relevant information is the overall 50 percent success rate. No matter the skill of the surgeon, the health of the patient, the severity of the ailment, there is still a 50 percent chance of survival, just like a flipping a coin.

However, if this were true, you should probably be very surprised to hear of your surgeon's recent record of success, which would then be exactly as likely as tossing a coin and getting heads 20 times in a row—roughly one in a million. It is certainly possible, but if you witnessed it happening, you would suspect that you do not have all the necessary information to understand why.

As Tom Stoppard writes in his play *Rosencrantz and Guildenstern Are Dead*, "Life is a gamble, at terrible odds. If it were a bet you wouldn't take it." The title characters in the play spend some of their limited lifetime flipping a coin and seeing it come up heads 92 times in a row. This is very unlikely indeed: 1/2 multiplied by itself 92 times, which is roughly 1 in 5 octillion, which is 5 followed by 27 zeros. But poor Rosencrantz and Guildenstern are doubly fictional, Stoppard having stolen them from *Hamlet*, and in neither of these fictional worlds do they possess the self-awareness or mathematical knowledge to be worried about their incredibly unlikely situation, nor their unfortunate fate.

So we should be more careful than they are. Perhaps the surgeon in our hypothetical scenario is indeed exceptionally skillful. Perhaps she usually takes referrals from patients with a very mild and easily treatable form of the ailment. Perhaps the hospital is in a wealthy area where the patients have access to excellent health care options, which aid in recovery. Conversely, perhaps most patients in need of this surgery have many other health issues, or have been exposed to a pathogen usually found in areas without easy access to high-quality health care. Perhaps the surgery has only been performed a small number of times, and the medical community only recently perfected the procedure.

Any of these possibilities—and many more—could result in a very skewed survival outcome, with half the overall patients dying from the surgery, but the patients of your surgeon having a high success rate. But you need to know which of these pieces of information applies. You need a model that you can trust. If you can get more data on why there is such a difference between the overall rate of success and that of your surgeon, you might be able to figure out how to weigh your options when evaluating how they apply to your specific case. Are you one of the patients who is like your surgeon's last 20 patients, who survived the surgery, or do you only have a 50 percent chance of surviving?

Since all probabilities are conditional, we should always be striving to get more information to improve our model.

So perhaps the non-mathematician mentioned earlier did understand this scenario better than the mathematician. Absent any other information at all, it is reasonable to assume that there is some nonrandom reason for your surgeon's otherwise unlikely string of 20 successes and that your odds really are better than the overall 50 percent rate. Both pieces of information apply in determining the conditional probability.

Once we've got our conditional probability, we can revisit our rules for combining probabilities. At the start of this chapter, we had a rule that let us simply multiply probabilities, but only when the two statements didn't depend on each other in any way, such as two rolls of a die, or two flips of a coin.

But for lots of statements, this sort of independence does not hold. The probability of a Democrat winning an election in New York might not be independent of the probability of a Democrat winning in nearby New Jersey. The probability of rain tomorrow might not be independent of rain today.

Using conditional probability, the probability of it raining both today and tomorrow is the product of the probability that it will rain tomorrow, given that it has rained today, multiplied by the probability that it will rain today, and all given some particular weather model. Our model—which could be our understanding of the local weather after a few years living somewhere, or which could be a massive computer simulation run by the UK Met Office or the US National Center for Atmospheric Research—might encode an enhanced probability of rain two days in a row, and this new formulation that includes conditional probability takes advantage of that.

In symbols, this scenario would be written

$$P(\text{rain today \& rain tomorrow} \mid \text{weather model})$$
$$= P(\text{rain tomorrow} \mid \text{rain today \& weather model})$$
$$\times\ P(\text{rain today} \mid \text{weather model}).$$

For example, we might expect it to rain only 5 days in November, so the probability of rain, $P(\text{rain today} \mid \text{model})$, is 5/30 or 1/6—the same as rolling a 1 on a die. Naively, we might then expect the probability of rain both today and tomorrow to be just like the probability of rolling a pair of ones: $1/6 \times 1/6$ or 1/36. But we should really be using our conditional probabilities to make our model. If the model predicts that consecutive days of rain are more probable than intermittent days of rain, we might have the conditional probability of rain tomorrow given rain today, assuming this more detailed model, is, say, 1/2 instead of 1/6, and so the probability of rain both today and tomorrow is $1/2 \times 1/6$ or 1/12, quite a bit more probable than 1/36.

Of course we can write this rule in our more general form, which is applicable to any pair of events, let's call them A and B, whose probabilities are linked by a model, M:

$$P(A\ \&\ B \mid M) = P(A \mid B\ \&\ M) \times P(B \mid M).$$

That is, the joint probability of the two events depends on the probability of one event given the other one, both being dependent upon some model. (Compare this to the previous equation, with A = "rain today," B = "rain tomorrow," and M = "weather model.")

We've got two slightly different kinds of conditional probability that we are working with here. First, all these probabilities are conditional upon the overall model, M, that we are considering. That might be a giant computer model of the weather over the UK or the USA, or it might just be the statement that I know that I am flipping a fair coin.

But sometimes, individual statements will be conditional upon each other, exactly because the model says that they should be.

Earlier I discussed the probability for the possibility that it will rain tomorrow and that I will forget my keys. I assumed that these two events were independent, so to determine their likelihood, I could just multiply the probabilities of each:

$$P(\text{rain and forgotten keys}) = P(\text{rain}) \times P(\text{forgotten keys})$$
$$= 0.30 \times 0.10 = 0.03.$$

But with the idea of conditional probability, we can be more careful in our model. What if I am often more flustered when I leave the house on rainy days, worried about my umbrella, my raincoat, the bus schedule, whether the Gore-Tex on my shoes has worn out, and other things. So perhaps rather than an overall 10 percent chance of forgetting my keys on rainy days, I instead have

$$P(\text{forgotten keys} \mid \text{rain}) = 0.20.$$

To incorporate this information into the model, I can use the conditional probability rule:

$$P(\text{rain \& forgotten keys}) = P(\text{forgotten keys} \mid \text{rain}) \times P(\text{rain})$$
$$= 0.20 \times 0.30 = 0.06.$$

Because I'm twice as likely to forget my keys on a rainy day than on a clear day, it's twice as likely for both conditions to happen, even if on average I only forget my keys 10 percent of the time and it only rains 30 percent of the time.[5]

What Is Probability?

We have seen that the rules I've mentioned embody some of the common-sense notions of probability and of how the probabilities of different statements are related to one another. These rules, sometimes known as the overly grand-sounding "calculus of probabilities," form a coherent mathematical theory, and its theorems and equations are proved in the same way as in the work of Pythagoras and Euclid. The mathematics of probability can be derived from yet simpler definitions or axioms in several ways.

In fact, the different derivations correspond to different ways of thinking about probability. They all give the same mathematical rules, so can rightly be called derivations of the same theory of probability, but they match with the interpretation of mathematical and numerical statements of probability in different ways.

Probability as a Measurement of Belief

The understanding that probability describes the strength of our beliefs goes back at least as far as Hume and his contemporaries such as Thomas Bayes and Richard Price, whom we will meet soon. But an actual derivation of the mathematical rules that maps most closely onto this interpretation of probability starts with Aristotle's logic of true and false statements augmented with axioms that should apply to any mathematical treatment of our beliefs. This line of thought began with Richard Cox in the 1940s and was strengthened by Ed Jaynes and Kevin van Horn in the late twentieth and early twenty-first centuries.[6]

Cox started from a small number of requirements for a mathematical treatment of "plausibility," purposely choosing a word other than "probability" before he presented his derivation. He shows that these requirements mathematically imply the rules of probability, defined as functions of statements that we want to make about the world—our usual candidates such as "it will rain tomorrow" or "I will roll a 3 on this die" or "a Republican will win the next US presidential election." Cox's derivation shows that we can assign probabilities between zero and one to a statement, and that they must obey the mathematical rules of probability that we have already seen: zero is true; one is false; we add probabilities when we want to represent "or" and we multiply them to represent "and." This derivation relies on conditional probabilities from the outset and highlights our slogan that *all* probabilities are conditional.

But Cox's derivation doesn't give us a guide to *using* these probabilities. What does it mean when I assign probability 0.9 to one statement and probability 0.45 to another? What does it mean that I am twice as sure of the first statement than the second?

Probability as Betting Strategy

To give an operational and practical description for the use of probabilities, we can use the so-called Dutch Book argument. "Book," here, has the same root as "bookie," and "Dutch" refers to the early twentieth-century mobster and gambler Dutch Schultz (né Arthur Flegenheimer)—much of probability theory comes from its applications to gambling.[7] This mathematical argument makes the case that you must define the betting odds that you would be willing to take on any outcome from the probability that you have assigned to that outcome. Otherwise, a clever bookie can arrange a series of wagers that will guarantee a loss for you, no matter the outcome. This is not a statement about whether your probability is a "good" one, considering all the known facts—perhaps you are 90 percent certain that a horse will win a race, even though it has a broken leg. The Dutch Book argument says that if you are willing to take 9 to 1 odds that this horse will win, you must also be willing to take 1 in 9 odds that it will lose. If your odds are not correctly related to each other (i.e., if they do not obey the rules of probability, for example if your probabilities don't add up to one), the clever bookie could arrange a set of bets that you would be certain to win or lose, no matter the outcome of the race.

This is exactly why a casino doesn't pay out based on the actual probability of a given bet, but takes a small cut for itself as guaranteed profit. For example, if we can apply the principle of indifference to the 37 different possible outcomes of a roulette wheel, we would assign a probability of 1/37 to each of them. But the casino only pays out 35 to 1—if you bet a dollar on each of the numbers, you would be guaranteed to win one of them and be guaranteed to lose money nonetheless, having spent $37 but won only $35.

This certain loss or gain is, in other contexts, known as arbitrage: taking advantage of incompatible assignments of prices or exchange rates of some commodity or currency to be sure of profit, no matter the fundamentals of the underlying investment.

An early proponent of these arguments was Bruno de Finetti, whose slogan "probability does not exist" we have already encountered. By aligning probability with betting preferences, he emphasizes that it only makes sense to talk about the probability assigned by an individual, or a consortium of like-minded individuals willing to agree on their betting odds.

So now we have a bit more understanding about the meaning of probability: if I believe something is more probable, I should be willing to bet more on its outcome, with odds precisely defined by the numerical value of the probability. Or perhaps it's the other way around: the odds we would be willing to take tell us what probability to assign. The logic of this second statement is often used to "elicit" the probabilities that people hold, or act as if they do, and by averaging over many people's willingness to bet on the outcomes of events, we can get some information on their overall likelihood given the information available and the models that people are considering. In this way, betting markets are often good predictors of the probability of the outcomes of sporting events or even elections.[8]

Probabilities as Frequencies

One of the first scientists to discuss probability in mathematical depth was Pierre-Simon Laplace. Laplace wrote *The Analytic Theory of Probability* in 1812 and *A Philosophical Essay on Probabilities* in 1814. He, too, largely based his mathematics on the principle of indifference. From this, he was able to derive the mathematical laws of probability.

Laplace certainly understood probabilities to describe a user's beliefs about the world, but he also acknowledged that one of the main ways that we confirm that our probability assignments make sense is through simple counting. If I assign probability 1/2 to heads in a coin flip, then I should expect roughly half of the flips to come up heads, getting closer and closer to 50 percent the more times I flip the coin. This is known as the law of large numbers: the probability of an outcome is equal to the fraction of times that I would observe some outcome (or find some statement to be true) if I repeated the observation many times.

The rough idea of the law of large numbers had been understood as the basis for gambling and games of chance for a long time. It was described mathematically by Gerolamo Cardano in the sixteenth century, with a first rigorous proof by Jacob Bernoulli, with embellishments and improvements by Abraham de Moivre and Simeon Denis Poisson, all in the eighteenth century.

During the nineteenth century, many mathematicians, philosophers, and scientists reified this observation into a *definition* of probability: the probability of an event is exactly the fraction that I would observe "in the long run."[9] This contrasts with our conditional definition of probability, which the users assign to some statements, based on models and data. But this so-called frequentist derivation allowed its champions to get rid of the underlying subjectivity of probability in favor of something that they took to be more objective and observable.

Probability as Mathematical Rules

Mathematicians and physicists might be more familiar with the very formal description of probability given by the Russian mathematician and physicist Andrey Kolmogorov in the 1930s. This derivation defines probability as a function of sets of possible "events"—although the name is suggestive, the derivation doesn't need to define what those events mean in the real world. We typically think of them as possible states of the world—but in fact Kolmogorov's definition is completely abstract. You can define probabilities over any such set, and a probability is just *any* set of numbers assigned to those events that obey the laws of probability. All these derivations, as a measure of belief, as a guide to betting, as frequencies, end at the same set of mathematical rules.

These different derivations are often thought to conflict with one another, representing mutually exclusive perspectives on what probability might mean. I prefer to think of them as different views of the same concept, each useful in its own sphere. Cox's derivation tells us that probabilities represent how strongly we should believe something to be true, based on other information. The Dutch Book argument gives us at least some guidance about how to act on those beliefs, and Kolmogorov's rule gives us further guidance on how to do the necessary underlying mathematics, especially useful when we apply the tools of probability to real-world scenarios. And finally, the frequentist interpretation of probability lets us loop back to some of our original notions of probability, linked to how often different kinds of events actually happen.

Bruno de Finetti, who championed the Dutch Book derivations in his own early work, came up with a remarkable theorem describing how the derivations and their different interpretations are related. He showed that the law of large numbers, and its predicted relationship between long-run frequencies and probabilities, could be understood entirely in the context of the subjective assignment of conditional probabilities. The theorem applies when we have a set of several outcomes that we want to consider together.

We might think that we must specify the probability for every possible outcome. For example, if we have a set of four results of a yes-no question, then we would need to separately specify the probability for each of the outcomes: YYYY, NYYY, YNYY, YYNY, YYYN, NNYY, NYNY, YNNY, YNYN, YYNN, NYYN, NNNY, NNYN, NYNN, YNNN, NNNN.

De Finetti's famous "representation theorem" defines a simple property that we might assume as part of our model in a variety of circumstances; this property is a symmetry that is usually called "exchangeability," when we have no reason to distinguish between different orders that those outcomes could have occurred in. (The word "exchangeability" comes from the slightly more specific definition that we could *exchange* any two of the outcomes and not be able to tell the difference.) For our example of the yes-no question, exchangeability implies that we would assign the *same* probability to each of outcomes YYNN, YNYN, NYNY, NYYN, YNNY, NNYY: the probability only depends on the fraction of Y and N in the outcome. That is, exchangeability lets us think of the model *as if* there is a probability assigned to the frequency of the answers yes or no in the outcome.

This is a kind of symmetry that often makes sense for rolls of a die, or flips of a coin, or marks on an exam (when, for example, we don't have information about the individual students). Of course, it doesn't always hold: the U.S. presidential election in 2020 was very different from the one in 2016 and certainly from the one in 1896. But when it does hold, it vastly simplifies the accounting and the models that are allowed given the assumption of exchangeability.

This representation theorem, and others that have followed it, allow us to assign probabilities based on our perceived symmetries of the problem at hand, and then treat the problem as if the outcome is controlled by an underlying probability, when (following de Finetti's dictum that "probability does not exist") this probability is really a consequence of our model.

As Walt Whitman wrote in *Song of Myself,*

> I am large, I contain multitudes.

I don't know about you, but I don't have a single set of beliefs about the world, and they certainly don't change only when I learn something new.

My opinions—my models of the world—can change depending on my mood, what books I've read, who I've spoken to recently.

But if we stick with our slogan that all probabilities are conditional—dependent upon some model—we can assign probabilities given any model at all. So I can assign the probability of a coin falling on heads given a two-headed coin (probability of one, since it's certain in this case), or a two-tailed coin (probability of zero, since it will never happen), or a fair coin (1/2, since heads and tails are equally likely). If I am playing poker, I could model my opponents' cards (and playing styles) in any number of ways, and each would give me different sets of bets that would be sensible to make, each self-consistent and externally consistent with the different models.

As a scientist, I could model the results of an experiment based on different theories and analyze the results under each. We might call these "as-if" probabilities, sometimes more technically known as "counterfactuals." These are the probabilities that we would assign to a situation if we believed the world, and the possibilities presented to us, to be a certain way. If two scientists agree on how they would assign probabilities in different situations, then they can reason together. The scientists don't have to discuss *their* actual beliefs at all.

This is objectivity, or at least it seems like the kind of objectivity worth having: we come to the same conclusions if we start from the same premises. Conversely, a single scientist—or the authors of a single science paper—could interpret the results of an experiment assuming different models and present them all to their colleagues.

We need to return to the concept of induction and show how scientists can manipulate probabilities, combining their models and their data to make inferences.

FOUR

"Our Deceased Friend Mr Bayes"

One of the main tasks of science is to help us augment our models with new information. That is, we want to learn from experiments and observations. The rules of probability help us do this. But they are mathematical rules, and we are going to need to use math to show this.

Remembering that all our probabilities are conditional, we want to be able to compare some hypothesized models, given both some new data and some information common to all the hypotheses. This information comprises an overarching background model, allowing us to compare different hypotheses to explain the data. In many cases, the hypotheses are very closely related, just differing by the value of some numerical parameter, and the background information encodes this.

Let's say we want to know the probability of our hypothesis, given both the data and the background information. In our mathematical notation, we write this as

$$P(\text{hypothesis} \mid \text{data \& background}).$$

We can get to this probability by starting with the slightly different quantity

$$P(\text{hypothesis \& data} \mid \text{background}).$$

Using the mathematical rules for manipulating probabilities—in particular, the "and" rule for multiplying conditional probabilities—tells us that $P(\text{hypothesis \& data} \mid \text{background})$ is equal to *both*

$$P(\text{data} \mid \text{hypothesis \& background}) \times P(\text{hypothesis} \mid \text{background})$$

and

$$P(\text{hypothesis} \mid \text{data \& background}) \times P(\text{data} \mid \text{background}).$$

These expressions are the same because the truth or falsehood of "hypothesis & data" is the same as for "data & hypothesis" — the order of those terms doesn't matter. Since they are equal to each other, we can use very simple algebra to solve for the quantity that we are looking for:

$$P(\text{hypothesis} \mid \text{data \& background})$$
$$= \frac{P(\text{hypothesis} \mid \text{background}) \times P(\text{data} \mid \text{hypothesis \& background})}{P(\text{data} \mid \text{background})}$$

Or, slightly more concisely, using H for "hypothesis," D for "data" and I for "background information,"

$$P(H|DI) = \frac{P(H|I) \times P(D|HI)}{P(D|I)}$$

This is Bayes' theorem, named after the Reverend Thomas Bayes (1701–1761), an English thinker and mathematician.

Bayes' theorem is just a mathematical fact, a consequence of the rules of probability, and is therefore true for any statements H, D, or I. But we can understand it a bit better if we are more precise in our interpretation of these quantities. The background information (I) appears to the right of the "given" or "conditional" bar in all our probabilities and is the model from which all our probabilities and inferences will flow. It describes our over-arching understanding of the questions we are trying to answer: What are the possible hypotheses (H) among which we want to select, and how does the data (D) follow from those different possibilities?

Each of these individual probabilities in Bayes' theorem has a name and a useful interpretation.

- $P(H|DI)$ is the *posterior*. This gives us the probability of our hypothesis given all the information that we have at hand, both the data and the background information.

- $P(H|I)$ is the *prior*. This represents the information that we had about our hypothesis absent ("prior to") the data. The prior can represent what we've learned from other experiments and information from discussions with others, as well as prejudices and biases that we, knowingly or otherwise, bring.
- $P(D|HI)$ is the *likelihood*. This is the probability of observing the actual data, D, if the hypothesis, H, is true. So this might be one of our counterfactuals again: we want to know the probability of getting any possible data, under the assumptions of our hypothesis, although eventually we will only need the probability for the actual data that we observed.
- $P(D|I)$ is often called the *evidence*. This represents how probable the data would be under the assumptions of our background information, but not the hypothesis itself. Because the model does not appear, the evidence does not give us any new information about the model (although in some circumstances we can use it to help judge the background information itself).

Considering this information, we can see a little more clearly why we needed our background information: since (according to our slogan) all probabilities are conditional, the background information is needed for all these probabilities. For the prior, in particular, we need a model at least to inform what we know about the hypothesis before encountering the data.

We can compress this by letting ourselves be a little sloppy and taking into account that the *evidence* doesn't depend on the model. Bayes' theorem can then become the even simpler

$$\text{Posterior} = \text{Prior} \times \text{Likelihood.}[1]$$

In other words: we update our *prior* information about our hypothesis by multiplying it by the *likelihood of our observed data given that hypothesis*. (The only effect of the *evidence* is to make sure that the posterior probabilities for an exhaustive set of models must add up to one. We can always work this out after the fact from the priors and likelihoods for all our models.)

So Bayes' theorem describes learning from data. It tells us in mathematical language how to revise what we know about a hypothesis, H, in the light of new data, D. But we can then use that probability as the background information when we get even more data.

Perhaps an apt description of this theorem would be something John Maynard Keynes is reported to have said (but probably never did): "When the facts change, I change my mind. What do you do?" Probabilities for the models and hypotheses in your mind change as you acquire more and more data—"facts."

Bayes' theorem is the mathematical expression of Keynes's aphorism. And this is exactly how we learn in practice, especially scientifically: if we measure the mass of the proton in a new experiment, we don't want to consider the result on its own. Rather, we want to combine the new information about the mass with everything that we've learned before.

More broadly, Bayes (and Keynes) tells us that we may need to change our models when the data no longer support them. Eddington, as we saw, showed that we should prefer Einstein's theory over Newton's when the measurements of gravitational lensing were closer to the predictions of general relativity.

This is part of how we link back to Hume, and how probability helps us justify induction, which is after all just a fancy name for the general process of learning from experience.

Bayes' Theorem: In Practice

I was recently visiting the Scottish capital, Edinburgh, home of David Hume, to work with my colleagues at the Royal Observatory and the university there. Looking at a map of the campus, I noticed Thomas Bayes Avenue. I knew that Bayes was English, from London, and I was surprised to see his name on a road in Scotland. But having seen his name on the map, I had to revise my opinion: surely, he must have some connection with Edinburgh.

This sequence of thoughts is Bayes's own theorem in action, albeit without mathematical precision: *prior* to seeing his name on an Edinburgh map, but knowing a little of his life, I would have bet against his having a strong connection with Scotland. Having seen the map, I had to consider the high *likelihood* that someone would need a connection to Edinburgh to get their name on a street there, overwhelming my prior skepticism. And indeed my *posterior* suspicion turns out to be correct: Bayes attended the University of Edinburgh, studying logic and theology, from 1719 to 1721, before returning to London.

Now that we have an actual mathematical theorem to guide us, we can be a little more precise in formulating our models and the probabilities that flow from them.

Postpandemic, we have become accustomed to thinking about the usefulness of disease testing and have realized that the presence or absence of a red line on a test does not tell us everything that we need to know. Bayes' theorem is exactly the tool needed to interpret the results.

We want to work out the probabilities of two different hypotheses: that I've got COVID (or whatever we are testing for; let's call this *hypothesis* "disease" to be more general), or that I am healthy. The *data* in this case is that I have tested positive, so let's call this condition "pos." Finally, we'll call the rest of our information about the problem *I*, encompassing what we know about the disease itself and the efficacy of the tests. Now, we can use Bayes' theorem to work out the probability that I've got the disease, given that I've tested positive:

$$P(\text{disease} \mid \text{pos} \ \& \ I)$$
$$= \frac{P(\text{disease} \mid I) \times P(\text{pos} \mid \text{disease} \ \& \ I)}{P(\text{disease} \mid I) \times P(\text{pos} \mid \text{disease} \ \& \ I) + P(\text{healthy} \mid I) \times P(\text{pos} \mid \text{healthy} \ \& \ I)}.$$

The numerator (top of the fraction) is the prior times the likelihood. The denominator (lower part of the fraction) is the evidence, which I've expanded to enforce the requirement that the probability of our two mutually exclusive models must add to one:

$$P(\text{disease} \mid \text{pos} \ \& \ I) + P(\text{healthy} \mid \text{pos} \ \& \ I) = 1.$$

Let's put in some realistic numbers.[2] A typical COVID lateral flow test might correctly detect the presence of the disease about 80 percent of the time. This reflects the *sensitivity* of the test, which in our probability language is the likelihood $P(\text{pos} \mid \text{disease} \& I) = 0.80$. It will incorrectly indicate disease in a healthy person about 2 percent of the time. This is the *false positive rate*, which translates to the other needed likelihood, $P(\text{pos} \mid \text{healthy} \& I) = 0.02$. Let us further assume a prevalence of about 5 percent for the people who are infectious at the time of the tests, giving the priors $P(\text{disease} \mid I) = 0.05$ as well as $P(\text{healthy} \mid I) = 1 - 0.05 = 0.95$. Putting these together gives

$$P(\text{disease} \mid \text{pos} \& I) = \frac{0.05 \times 0.80}{0.05 \times 0.80 + (1 - 0.05) \times 0.02} = 0.68.$$

Under these conditions, a person who tests positive has a 68 percent chance of having the disease, or about 2 in 3. This is less than the 80 percent sensitivity of the test, brought down by the low prevalence combined with the small, but nonzero, false positive failure rate for the test.

The probability of having the disease given a positive result (our posterior) is not equal to the probability of getting a positive result given the disease (the sensitivity). In words, this can be confusing, but in symbols, it is obvious:[3]

$$P(\text{disease} \mid \text{pos} \& I) \neq P(\text{pos} \mid \text{disease} \& I).$$

The posterior is not equal to the likelihood. The whole point of Bayes' theorem is to give the correct relationship between these two quantities.

Confusion between the likelihood and the posterior is a common logical fallacy. In Malcolm Gladwell's *Outliers*, he notes that many experts have had to put in roughly 10,000 hours of practice before becoming successful.[4] He applies this to the Beatles and their many hours practicing in Liverpool and Hamburg before becoming the greatest rock 'n' roll band of their era (and beyond, depending on your predilections), to Bill Gates's time spent programming computers before the success of Microsoft, to chess players, and more.

In his book Gladwell is usually careful to point out that this vast amount of practice is a necessary condition for success, but it is not sufficient. In our probability language:

$$P(10{,}000 \text{ hours} \mid \text{success}) \neq P(\text{success} \mid 10{,}000 \text{ hours}).$$

The left-hand side of the equation represents the probability that someone who is successful will have put in 10,000 hours of practice. This is somewhat empirically supported, at least for some areas of expertise (and if you take "10,000 hours" as a broad average). But the right-hand side might be more interesting to us less successful people: What is the probability of success if you put in 10,000 hours? Sadly, to relate these two quantities requires the full machinery of Bayes' theorem and restoring any needed further background information, I—which now encompasses Gladwell's research for his book and the sources upon which he drew:

$$P(\text{success} \mid 10{,}000 \text{ \& } I) = \frac{P(\text{success} \mid I) \times P(10{,}000 \mid \text{success \& } I)}{P(10{,}000 \mid I)}.$$

So even if the likelihood, $P(10{,}000 \mid \text{success \& } I)$, is very close to one (i.e., almost all successful people have worked 10,000 hours at becoming experts), the probability that any given individual will succeed even after putting in their time depends on the prior probability of success, $P(\text{success} \mid I)$, which can depend on innate talent, not to mention luck.[5]

How to Make a Measurement

Making measurements is where models and probability overlap.

So far, we've talked about probabilities as we've been using them to describe our knowledge about some statement, such as determining whether someone has a disease after seeing their test results. With Bayes' theorem, we saw how we could calculate the posterior probability, $P(H|DI)$, the probability of some hypothesis, H, given data, D, and the ubiquitous, and always necessary, background information, I.

But most of science is more complicated than testing the truth of a single statement. Mostly, we scientists make measurements. What is a measurement? A measurement is just an attempt to figure out the value of some number—sometimes called a *parameter*—by doing an experiment or observation. This can be as fundamental as trying to determine properties like the mass of some elementary particle such as an electron, or the distance to a star, or even something much more prosaic like the length of some piece of equipment in a laboratory.

How do we fit these measurements into our framework of probabilities? What are the models that we are trying to compare? We somehow need to express the idea that I don't know exactly the value of some quantity that I'm interested in. Now, the competing hypotheses (H) are all the different possible values of the parameter, and the background information (I) is my knowledge of the measurement process and anything I already knew about the value before making the measurement. The data D and background information I are fixed, and there are many different possible values of the hypotheses H. Sometimes we use a small letter like h or even the Greek θ (theta), which for the mathematically inclined connotes something that can take on a numerical value.

The simplest example of using this information to make a measurement is something that we've all done: using a ruler to measure a length. Now, the competing models are all the possible different lengths that it might have. It could be an inch long, or a mile. Or 1.371652231 inches. Or 0.00000012 inches. . . . If all I know is that I am measuring a length, then about the only thing I know is that the answer is a number greater than zero. So a priori, the different hypotheses are all the possible lengths: a continuous measurement, choosing among all the possible real numbers with any number of decimal places. That is, our different hypotheses or models are the infinitude of all the possible numbers! We can express this not by listing the probabilities for the infinite number of possible lengths, but by a mathematical function that gives the probabilities for any height at all.

My prior probability might simply be that any possible length is equally likely—using a version of the principle of indifference from Laplace and Keynes. Let's say I am measuring the height of my seven-year-old daughter. My background information might just be that she is a human being, and so might be around 30 centimeters tall, or 31 centimeters, all the way up to, say, 230 centimeters. Of course, those very short and very tall heights are not particularly realistic if I have any idea of her age, but we are free to use such counterfactual as-if probabilities to consider a model where I don't know

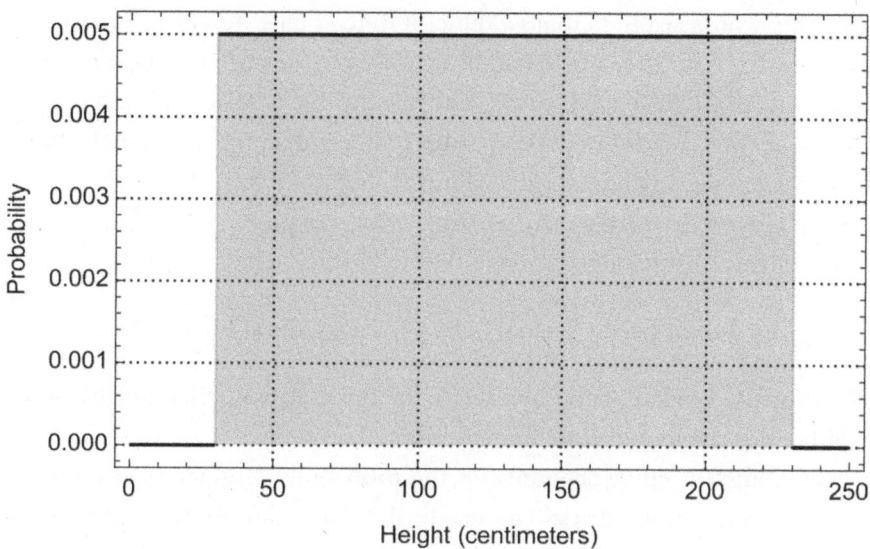

Figure 3. The uniform *prior probability* for my daughter's height, with equal probability over the range 30 to 230 centimeters, and zero probability outside of that range.

enough about her to make any distinctions before making the measurement. So very roughly we can say that the prior, P(height | background) and shown in Figure 3, is a straight horizontal line, a constant, independent of the height over the range of 30 to 230 centimeters.

I also need to know the likelihood: the probability of getting a particular measurement given her actual height (which, remember, I do not actually know!). How does this measurement process work? If I'm measuring my daughter's height, I usually try to stand her next to a wall, lean a book on her head, and scratch a line on the wall where the book touches it. If you've ever done this with a squirmy child and a wonky old house, you know that this is not a very precise way of measuring a kid's height. We know that the measurement taken from this process is somewhere near the right answer, but there is a reasonable chance that it's off by at least a small amount—with the probability of the error decreasing the further you get from the right answer. So we can describe the probability as a curve, peaked at the actual height, and getting smaller as you get further away, with a width given by a number that describes how much I trust the full measurement process: the squirmy child, the wonky walls, perhaps a difficult-to-read tape measure. So let's say the curve has a width of about half of a centimeter or so. (The shape is often described as a bell-shaped curve, and has the more technical name of a Gaussian, after Carl Friederich Gauss, an eighteenth- and nineteenth-century German mathematician who first described it in detail, then rapidly picked up by Laplace in his own work on probability.) This curve describes our likelihood, the probability

P (measured height | actual height & background information).

The measured height won't be exactly the same as the actual height—the likelihood is not identical to the posterior.

The Gaussian curve accounts for the probability of making an error in the measurement, but it is still an idealization of the measurement process. A Gaussian curve never gets to zero: there is always some probability, albeit very small, of making an arbitrary large error in the measurement. This is not realistic: despite the squirming child and wonky walls, nothing would make me read "1.23 miles" for my daughter's height, much less "−67 centimeters," which is just physically not allowed, since all heights are positive! (On the other hand, if I allowed the five-year-old to measure the height of the seven-year-old, perhaps both are at least marginally possible!) Figure 4 shows the likelihood as a function of the measured value if the unknown actual value were 103 centimeters. In this case we would expect to read a number near 103 centimeters off of our ruler, with appreciable probability of seeing 100 centimeters, say, or 105 centimeters, but with much less probability of seeing 95 centimeters, or 125 centimeters. (This probability is hard to see in the figure, but remember that the curve *never* goes all the way down to zero.)

Acknowledging these idealizations, and armed with the prior and the likelihood, we have everything that we need to work out the posterior: If I read off, say, 110 centimeters, what can I say about my daughter's actual height? Well, I have to multiply the prior by the likelihood:

$$P(\text{height} \,|\, 110\,\text{cm} \,\&\, I) = \frac{P(\text{height} \,|\, I) \times P(110\,\text{cm} \,|\, \text{height} \,\&\, I)}{P(110\,\text{cm} \,|\, I)}.$$

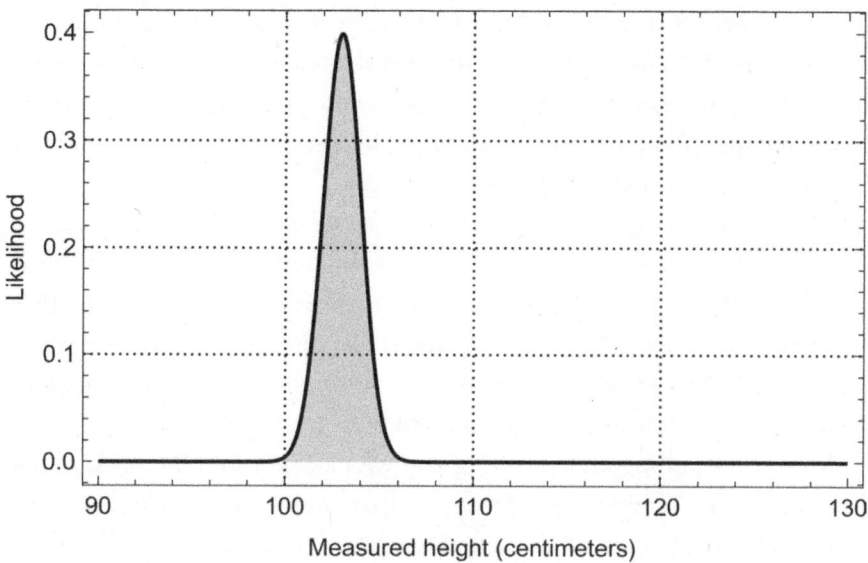

Figure 4. Bell-shaped curve for the *likelihood*, *P*(measured height | actual height) as a function of the *measured* height, if the *actual* height were 103 centimeters. In a measurement, the actual height will always be unknown, but this curve can be calculated for any possible value that it might take.

Here "110 cm" stands for the reading of the ruler, and *I* is shorthand for my background information as before. So in this case, the posterior is simple: my prior is a constant across all the possible heights, and the likelihood is our bell-shaped curve, so the posterior has the same Gaussian bell shape. But Bayes' theorem has allowed us to pull a neat trick with our probabilities. Our likelihood was defined as the probability of getting a particular *measurement* given the *actual* height. We now use the expression for the likelihood in the opposite way: fix the measurement to whatever number we read off the ruler, and let the *actual* height vary. We also have to multiply by the prior, but in this case the prior is a constant, and so doesn't change the shape of the curve, except that it makes it go to precisely zero for heights less than 30 centimeters and more than 230 centimeters.

We have made the—slightly unrealistic and idealized—assumption that the accuracy of our measurement is the same no matter the actual value of the height. (For example, very large values of the height might be larger than our ruler and make it difficult for us to perform the measurement.) In this case, the likelihood looks the same whether we let the actual height vary, or the measurement: what started out as a bell-shaped curve as we changed the number that we read off our ruler becomes a bell-shaped curve as we change our hypothesis about my daughter's actual height. Now, Figure 5 shows the posterior probability curve for a measurement of 110 centimeters.

This, finally, is our *inference* for the actual height of my daughter: it too is described by a bell-shaped curve, with the highest probability at the same number as the measured value, with a width given by the width of the likelihood curve, or about half a centimeter in this case.[6] It is much more probable that she is 109 centimeters than 97 centimeters or 123 centimeters. The width of the curve is often described as an *error bar*.

It's worth dwelling on everything that we have put into the background information, *I*. We needed to specify that there was a set of models, each described by a single value of the actual height. We needed to describe the relationship of our measurement apparatus—the tape measure, wall, pencil mark, and perhaps a book to help us line up the top of my daughter's head with a place on the wall, and we needed to use this information to describe any error associated with the measurement. All these together generate the likelihood, the probability of reading out a particular value on the ruler for each of the possible actual heights of my daughter. We also need to describe what we already knew about her height before we attempted the measurement. But we are free for this description to encode the completely unrealistic situation in which we don't know anything about her height: she is as likely to be between 20 and 30 centimeters as between 110 and 120 centimeters. This is unrealistic—I know a lot more than this!—but perfectly reasonable as a starting point.

However, I'm *also* free to consider a more informative background model in which I am pretty sure that she is somewhere near 100 centimeters tall, for example, based on previous measurements (or just looking at her!). But only when the prior information becomes comparably restrictive to the measurement on its own (that is, the likelihood) do the details of the prior matter. Bayes' theorem lets us combine arbitrary pieces of information about our hypotheses. If my spouse measured our daughter's height this morning, and I tried again this evening, using the same tape measure and

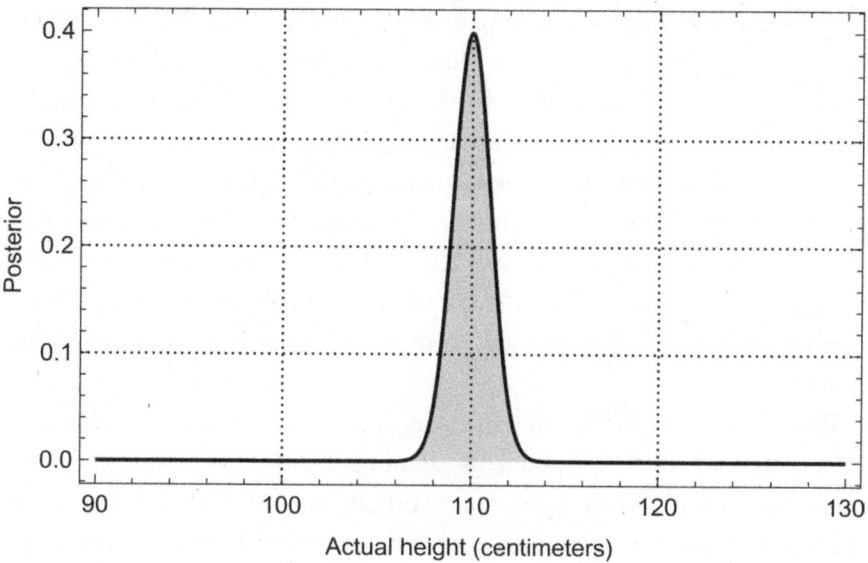

Figure 5. Bell-shaped curve for the *posterior probability* of my daughter's height, P(actual height | measured height & I) as a function of the *actual* height, if we make a measurement of 110 centimeters.

against the same wonky wall, we can just average the two measurements, although if there are any differences between the two "experiments" (different tape measures, varying degrees of squirminess) we must be more careful. And if we waited a few months in between measurements, we'd have to consider her growth during that time.

Leaving the example of my house and family for a moment, let's say I want to measure the distance from Earth to some star. This is quite hard to do, even compared to the heights of squirming kids standing against wonky walls, since we can't use any of the techniques from our everyday life for getting distances: I don't have any ruler long enough, and I certainly can't get to the star to put one end of a measuring tape there and keep the other end here on Earth. Even the nearest stars are so far away that even a signal traveling at the speed of light would take years to get there and back.

Instead, astronomers have to build up a set of techniques, first allowing us to measure distances from Earth to nearby stars in our galaxy in essentially the same way that having a pair of eyes allows us to judge distances—with parallax. With parallax, the slightly different pictures from our left and right eyes are combined by our brains into a three-dimensional view. But for distances to stars, we can use telescopes and satellites that essentially put the different views twice as far apart as the distance from Earth to the sun. We first learned to do this by comparing the pictures taken by terrestrial telescopes six months apart and then by using similar but even more accurate techniques from space telescopes such as the Gaia satellite. This lets us measure the distances to thousands of stars in our own Milky Way Galaxy.

We can measure the distances to stars even farther away (and to the galaxies within which they reside) by adapting another technique that our brains do naturally: if we know that two light bulbs are actually the same brightness (the same wattage) as each other, then we know that if one appears dimmer, it must be farther away. Similarly, if we know that some stars are the same intrinsic brightness (often called "standard candles"), dimmer ones must be farther away—and again, we can be more accurate by using precise astronomical measurements. Techniques like this can take us all the way out to the most distant objects that we can see—galaxies that are more than 20 billion light-years away.

Despite these spectacular distances, once we put all these different methods together, the processes are not so different from measuring my daughter's height. That is, I still need Bayes' theorem to combine what I already knew about the distance to a star (the prior) with the new information from the measurement (the likelihood) to give me my posterior probability. And as with measuring my daughter's height, that posterior of a star's distance is described by a curve depending on the details and complications of the measurement procedure, built up from nearby stars to the most distant galaxies.

In many cases, these will trace out the bell-shaped Gaussian curve: peaked around some particular distance with some width. When this is the case, or at least approximately the case, we can summarize the curve with just two numbers: the location of the peak and the width of the curve around the peak, as shown in Figure 6. For historical reasons as well as details of the

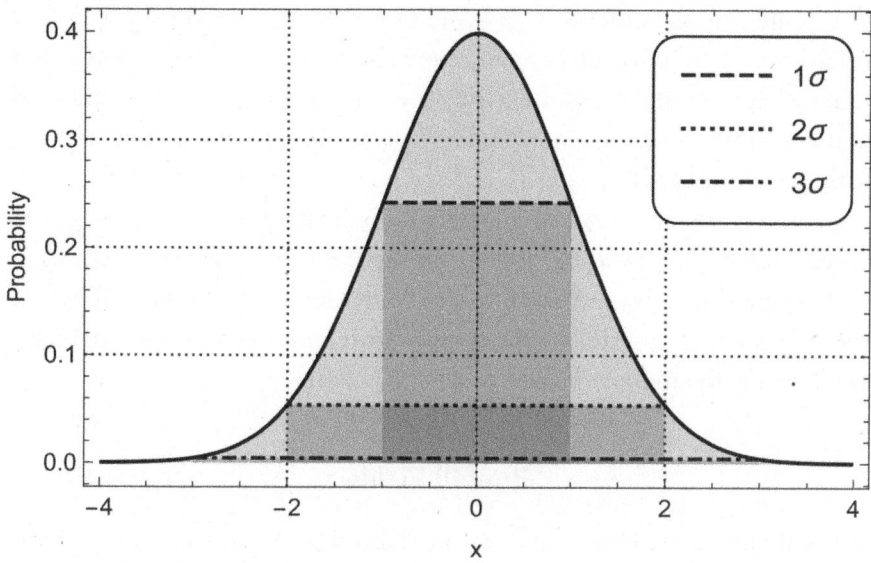

Figure 6. Bell-shaped Gaussian curve, with the peak centered at zero, with 1-σ, 2-σ, and 3-σ error bar widths.

Gaussian formula, we often specifically calculate the width as the interval, which includes 68 percent of the probability ($P = 0.68$), and this is used as the error bar for our measurement. The term "one sigma error," using the Greek letter σ, refers to this probability.

Using $P = 0.68$ for the error bar indicates that there is still about a one in three chance that the real value is outside of what is sometimes also called the "margin of error"—there is no hard cut-off, and we should not be any more surprised to find that the correct result is a little bit outside the error bar than a little bit inside it. For the Gaussian curve, going out to twice the size of the error bar (2σ) would increase the probability to about $P = 0.95$, or a 5 percent chance of being outside it, and tripling it to 3σ would raise it to $P = 0.99$, which means there is still a 1 percent chance of the real answer *not* being between those limits.

(Another way to look at these numbers uses the correspondences that we have seen between individual probabilities and long-run frequencies of repeated measurements, which indicate that we should expect to find results outside of those 68 percent errors about one-third of the time. That is, if we repeated the experiment 100 times, we would expect that about 68 of them would have the correct answer within 1σ error bar and 99 of them within 3σ error bars.)

And all these numbers assume that our model is a good description of the system that we are trying to measure and of the way that we are measuring it. If—unfortunately and realistically—our models are wrong, then we should expect to misestimate the actual errors and might find the right answer outside the margin of error even more often.

Bunhill Fields

Bayes' theorem, the basis of all these probabilities and measurements, stems from the work of the Reverend Thomas Bayes, who died in 1761, at just 59. He never got to see any of the repercussions of his most famous paper, "An Essay Towards Solving a Problem in the Doctrine of Chances," posthumously read to the Royal Society of London in 1763 by his younger friend and collaborator, Richard Price. (Figure 7 shows a timeline of the major figures and ideas discussed in this chapter.)

Bayes and Price are both interred in Bunhill Fields, a four-acre cemetery just outside the City of London, today nestled between the headquarters of banks and financial-services companies. In fact, Bayes's tomb is now maintained from afar partially by a hedge fund headquartered in New Jersey (USA), presumably as meager thanks for the financial insights that his formula has given them.

One of the other stones in the cemetery marks the resting place of John Bunyan, author of *Pilgrim's Progress*; another is for Daniel Defoe, who wrote *Robinson Crusoe*. But my favorite, along with those marking the graves of Bayes and Price, is the one for the poet and engraver William Blake. Blake is most famous for his poems that have become part of the canon of English literature: the hymn "Jerusalem,"[7] which is almost a second national anthem in England; "The Tyger," which most of us learned in school, with its famous not-quite-rhyming couplet "what immortal hand or eye / could

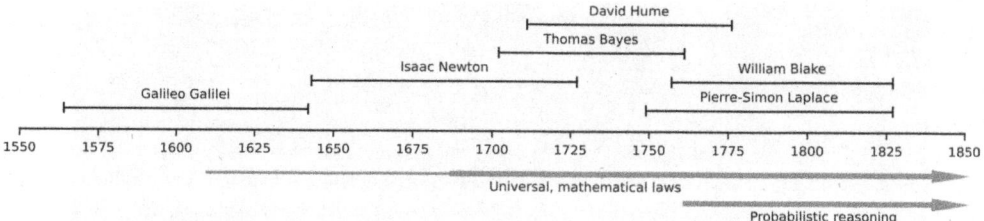

Figure 7. A timeline of the major figures discussed in this chapter, as well as the start of the eras in which (1) Galileo and Newton developed the ideas of mathematical models of science and (2) Bayes, Price, and Laplace promoted probabilistic reasoning with Bayes' theorem, both of which continue in modern scientific practice.

frame thy fearful symmetry," along with the rest of the short poems from his *Songs of Innocence* and *Songs of Experience*.

But his work was dominated by mythological, prophetic poetry, recounting his personal mystical-Christian cosmology. Blake lived at the dawn of the modern world,[8] at a time when Isaac Newton's ideas of a mechanistic Universe had only recently taken hold. Blake could not abide this model of the world. In a famous painting (Figure 8), he depicted Newton, muscular, naked, leaning over his geometric tools, calculating and configuring his world. To modern eyes, this lean and fit Newton looks like a superhero, and you can find postcards and posters of Blake's rendition of Newton pinned to the walls in physics departments everywhere.[9]

But to Blake, Newton was a false god, ignoring the natural beauty around him in favor of his geometry and calculations that purported to rule the behavior of the world and those in it. So Blake would perhaps not like the use to which the work of Bayes, his eternal companion, has been put by us scientists, further systematizing and modeling the world. "I must Create a System, or be enslav'd by another Man's. I will not Reason & Compare; my business is to Create." These are the words of Los, one of Blake's four Zoas—his invented mythology, recounted through several of his book-length poems, was complicated and not entirely consistent—but they surely echo his own rage against what he saw as the restrictive legacy of the Age of Reason and the deistic clockwork Universe that sprang up after Newton.

Figure 8. *Newton*, by William Blake (1795–1805). Color print with pen, ink, and watercolor. Tate Britain, London. (Wikimedia Commons.)

The First Bayesians

Richard Price's presentation of Bayes's paper began with an encomium, written in florid eighteenth-century prose:

> I now send you an essay which I have found among the papers of our deceased friend Mr Bayes, and which, in my opinion, has great merit, and well deserves to be preserved. . . .
>
> He had, you know, the honour of being a member of that illustrious Society, and was much esteemed by many in it as a very able mathematician. In an introduction which he has writ to this Essay, he says, that his design at first in thinking on the subject of it was, to find out a method by which we might judge concerning the probability that an event has to happen, in given circumstances, upon supposition that we know nothing concerning it

but that, under the same circumstances, it has happened a certain
number of times, and failed a certain other number of times.

Bayes's paper itself is very mathematical, but also written in the style of
eighteenth-century scientific works: many of the formulae are spelled out in
words, and much of the reasoning is geometrical, rather than algebraic, following in particular the example of Newton in his *Principia Mathematica*.
Price's posthumous introduction explicitly claims Bayes's result as a mathematical rule for justifying induction, albeit without a mention of Hume's
name.

Early on, Bayes defines his notion of probability:

> The probability of any event is the ratio between the value at
> which an expectation depending on the happening of the event
> ought to be computed and the value of the thing expected upon
> its happening.

With this definition, Bayes asks us to consider some number of observations
that might, or might not, satisfy some criterion, and seems to define probability as the ratio of the expected value to you if some event occurs to the total
payoff if it does occur. As with the Dutch Book argument, he starts in a context of gambling, but there is no need to stay there; the theory is applicable to
any binary choice. What is the probability that a voter will prefer one candidate over the other? What is the probability that it will rain on any given February day in London? We might have incomplete information from a poll or
weather records, and we want to find out the probability that some unpolled
individual will favor the candidate, or that it will rain next week.

A common scientific example brings us back to medical testing. Earlier
we asked what we could infer about having a disease given a positive test result and found that we needed to know both the sensitivity of the test and its
false positive rate—that is, we needed to know what fraction of the time we
would expect to see the result that we saw from the test, given having the disease, and given not having it. We took those fractions as a given, but where
do they come from? We need to measure them, too. This is mathematically
the same as the problem that Bayes considered 250 years ago.

We start from a group, all of whom we know to have the disease. We then administer the test and find that some number of them show a positive result. What does this number tell us about the actual sensitivity of the test? (We are assuming that we have some way, separate from the test whose properties we are assessing, for knowing whether someone has the disease; this could be a much more involved and invasive test, for example. We could build a much more complicated model that could describe the situation in which we don't have any way of attaining such information.) Each of those groups is described, separately, by Bayes's calculation.

For the polling example, we start with the total number of respondents, some smaller number of whom prefer our candidate. For the weather, we could start with historical records of some total number of past London February days and work out on how many it rained.

Regardless of the situation, Bayes assumed that each observation, past and future, could be described by a single probability, giving the chance that the test would come out positive if the patient has the disease, or that it will rain, or that one candidate is preferred. We might expect that the probability would just be given by the fraction of positive results in the example of the person who may or may not have a disease. But for the same reason that not every set of 10 coin tosses results in exactly 5 heads and 5 tails, the observed fraction and the underlying probability are not necessarily equal. Instead, we use Bayes' theorem to make inferences about that underlying probability, given the observed numbers.

This may seem a little strange: we are using Bayes' theorem to calculate the probability of a probability. Really, we are thinking of the sensitivity of the test for the disease, or the fraction of voters who will end up voting for our candidate, as a parameter that we are trying to measure. In the case of my daughter's height, the outcome of Bayes' theorem was a function, giving the conditional probability $P(\text{height} \mid \text{measurement})$, which ended up looking like a bell-shaped curve for the height, centered on the value that we read off our ruler. We are now looking for Bayes' theorem to give us a function

$$P(\text{sensitivity} \mid \text{number of positive and negative results}), \text{ or}$$
$$P(\text{prefers the candidate} \mid \text{poll numbers}).$$

Bayes himself sets out his problem as:

Given the number of times in which an unknown event has happened and failed: *Required* the chance that the probability of its
happening in a single trial lies somewhere between any two degrees of probability that can be named.

(In his "definitions," Bayes explicitly says "By *chance* I mean the same
as probability," so he understands that he is looking for the probability of a
probability.)

Bayes understood that his calculation required what we now call both a
prior and a likelihood. In fact, the notion of the prior does not explicitly
appear directly in Bayes's work—it took until the nineteenth century and
the work of Laplace before the modern form of what we now call Bayes' theorem, and the explicit split between prior and likelihood, was discussed. But
Bayes understood and derived the form of our "and" rule for the product of
conditional probabilities and how that would link the two together.

So Bayes was aware of the need for a prior, and he engineered a physical
system for which the prior seems obvious: he had us consider a surface,
such as a pool table, onto which we roll a ball. The ball could end up anywhere on the table, depending on the location, angle, and speed of our roll.
We then ask what fraction of subsequent rolls we should expect to end up
to the left or right of this initial location. In this case, Bayes claims, there is
no controversy about the prior probability for the first ball's location—it
could equally well be anywhere across the table—what we would now call
a "uniform probability." (In fact, what Bayes proved with his pool table
analogy was not quite what is needed to arrive at the prior probability for his
problem.)[10]

The likelihood is also easy to write down as a mathematical function using the rules of probability. Let's call the sensitivity of the medical test s and
the number of positive test results N, representing the people who do have
the disease, out of the total number of people who were tested T (so there
are $T-N$ negative results). If the patient has the disease, the chance of getting a positive result in a single run of the test is s, and the chance of getting
a negative result is $1-s$. We can use our "and" rule to combine these by just
multiplying together N copies of the positive probability s and $T-N$ copies

of the negative probability $1-s$. We then multiply this by our prior for s— which is just a flat line, and so doesn't have any effect—and that's it. So the shape of our posterior $P(s|T,N)$ is just given by

$$s^N \times (1-s)^{T-N}.$$

The same function would apply in our other examples, just substituting in different values for our variables: for the poll, s would be the chance that any given voter prefers our candidate, given that N out of T preferred her in the poll; for the weather, s is the chance of February rain, given a record of N out of T past rainy days.

Figure 9 shows the posterior for various values of N and T. We see that the probability for the value of s *does* peak at the observed fraction N/T and gets narrower and narrower as those numbers get larger.

This explanation is part of the link to—and the solution to—the problem of induction: the more observations we make, the surer we become.

Bayes could have just written down the formula above and stopped there, but in fact he had set himself a more difficult task: he wanted a formula for "the chance that the probability of its [an event's] happening in a single trial lies somewhere between any two degrees of probability." Just as with my measurement of heights and lengths and other quantities with error bars, this curve is really used to work out the probability enclosed between the points that it describes. The formula requires computing an integral, one of the then recent tools from Newton's calculus, at Bayes's time only a few decades old. Bayes was a mathematician of considerable ability, and much of his paper on the subject is concerned with how to best approximate the required numbers, since they cannot be written down as a simple formula and lacking today's computers were extremely laborious to calculate. These are heroic and sometimes mathematically insightful calculations, but for the modern reader this is the least interesting aspect of the paper.

Bayes's friend Richard Price has been called the "greatest Welsh thinker of all time" by the modern historian John Davies.[11] In addition to presenting Bayes's posthumous work to the Royal Society, Price worked out more of its repercussions and developed many of the mathematical principles of actuarial

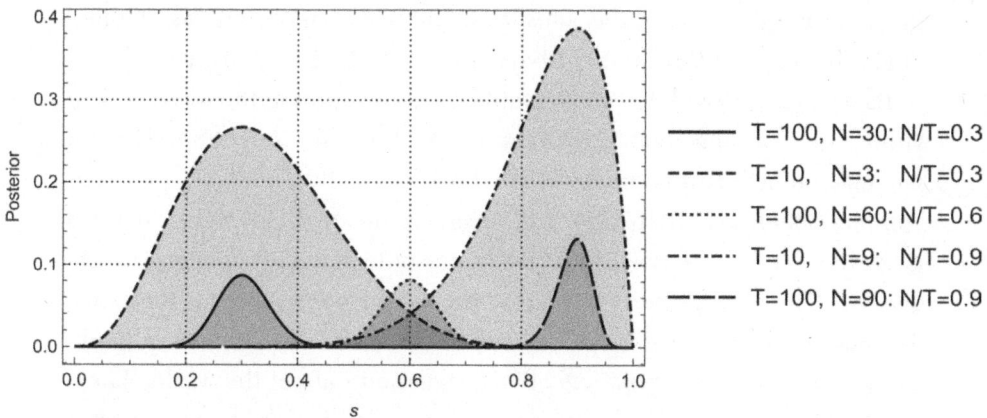

Figure 9. The probability for the quantity s (test sensitivity, voter preference, chance of rain) for different values of N observations out of T total trials.

science and insurance. He also supported feminist Mary Wollstonecraft as well as the American and French Revolutions.

Price also added a lengthy appendix to the paper. He first gives some numerical examples of Bayes's formulae. (Bayes was evidently more interested in the calculations themselves than in applying them to any particular case.) He goes on to show how we become more sure that the underlying probability is given by the observed ratio, as the total number of observations grows, until we can be "morally certain" that we know the outcome. He applies this to observations of 1 success in 10, 10 in 100, 100 in 1,000, and so on, showing that we become more and more certain that the probability is 10 percent.

He also considers the seemingly simplest case, where *no* successes are seen, reminding us that, even then, it is not certain that the underlying chance is zero, although again we become more sure of that chance as the total number of observations increases.

Price also goes beyond Bayes's discussion of the appropriate prior probability (although the term had not been invented yet). He realizes that some sort of observation is necessary just to give the basic information about the problem—the background information needed to really apply his friend's theorem. For this reason, the statistician and historian of statistics Stephen Stigler has called Price (and not Bayes himself) "the First Bayesian."[12]

Bayes came up with his ideas about probability around the same time as Hume was worrying about induction. In "An Enquiry Concerning Human Understanding," Hume discusses the difference between *relations of ideas* and *matters of fact*. By relations of ideas, he means those things that we know to be true by reason alone: the truths of logic, mathematics, and geometry. These things are necessarily true, as a consequence of their own definitions and the purely deductive connections that follow from them. These are contrasted to matters of fact, statements about the world. These statements may well be true, but we cannot prove them in the same way as relations of ideas. His most famous example is that we are sure that tomorrow will come: "That the sun will not rise to-morrow is no less intelligible a proposition, and implies no more contradiction than the affirmation, that it will rise." The very idea of cause and effect seems to be in the category of "matters of fact"—a contingent statement about the world. To discover a cause, we need to generalize from experience, but this leap cannot be logically ironclad.

In a different work, "A Treatise of Human Nature," he puts it most succinctly: "There can be no demonstrative arguments to prove, that those instances, of which we have had no experience, resemble those, of which we have had experience." Elsewhere in that work, he wrote of his worry that "all knowledge degenerates into probability"—here, he means to contrast "probability" unfavorably with "certainty," implying that probabilistic knowledge isn't really knowledge at all.

Later in "An Enquiry Concerning Human Understanding," he takes on probability directly as well: "Though there be no such thing as Chance in the world; our ignorance of the real cause of any event has the same influence on the understanding, and begets a like species of belief or opinion." This is a good slogan for what we now think of as the Bayesian view: Hume had an inkling that probability was the solution to his problem of induction. And the following quote is even better, a verbal description of Bayes's theorem without mentioning it:

> There is no Algebraist nor Mathematician so expert in his sci-
> ence, as to place entire confidence in any truth immediately upon
> his discovery of it, or regard it as any thing, but a mere probability.
> Every time he runs over his proofs, his confidence encreases; but
> still more by the approbation of his friends; and is rais'd to its ut-
> most perfection by the universal assent and applauses of the
> learned world. Now 'tis evident, that this gradual encrease of as-
> surance is nothing but the addition of new probabilities, and is
> deriv'd from the constant union of causes and effects, according to
> past experience and observation.[13]

But lacking the mathematical understanding that probability could be applied systematically, he ended up eschewing it as hopelessly subjective (as
have many other critics of Bayesian reasoning in the centuries since).

But where he really came into conflict with Bayes and Price was later in the
same essay, in the section "Of Miracles." Hume, who professed to be merely a
skeptical Christian, was thought by some to have come to what at the time was
dangerously close to proper atheism. Our only evidence of religious miracles
is simply hearsay, he wrote, communicated to us through the Bible or later accounts. Conversely, he wrote, "a miracle is a violation of the laws of nature;
and as a firm and unalterable experience has established these laws, the proof
against a miracle, from the very nature of the fact, is as entire as any argument
from experience can possibly be imagined." This, too, turns out to be Bayesian
reasoning *avant la lettre*: our prior that the laws of nature must hold always, in
all situations, is so strong that we should not trust these many-times-removed
eyewitness accounts of instances when these laws do not hold.

Bayes—who was, after all, a Presbyterian reverend—disagreed and constructed a counterargument. It may be true that we should initially be skeptical of a report of a miracle, but there have been many such reports, of many
different miracles. Bayes showed that you can mathematically pull yourself
up by your bootstraps: the more times you observe some phenomenon, the
more you should expect it to happen again. Really, Bayes and Hume were
using the same argument: both observations and nonobservations of some
phenomenon can make us surer of the underlying probabilities.

It has recently become clear that Bayes's essay, including Price's introduction and appendix, were intended as rebukes to Hume's argument in "Of Miracles."[14] Drafts of previous versions of Bayes's essay have been found, with the alternate title "A Method of Calculating the Exact Probability of All Conclusions Founded on Induction" (although Hume himself is not named).

Price went on to further elaborate Bayes's work, reading an essay to the Royal Society in 1764 intended as a supplement to Bayes's paper and then in 1767 publishing a book, *Four Dissertations*, which contained a section bringing together the different strands, "On the Importance of Christianity, Its Evidences, and the Objections Which Have Been Made to It." Here, Price deploys probability to argue against Hume. He returns to his numerical example of the repercussions of observing zero successes out of a large number of observations and applies it to the case of one million. We (humanity) have observed the high tide reappear regularly for, say, one million days (about 2,700 years)—its nonobservation even over this long period would have been so significant that we surely would have heard if it had not occurred! What is the probability—given the model of Bayes and Price—that we still might observe a day without a tide in the next million? With the mathematics Price had available, he could not answer this question exactly, but was able to show that there was indeed a significant probability. It would take the work of Laplace in the ensuing decades to further systematize probability to the point that this question could be answered more precisely. Laplace used much of the same mathematics as Bayes did, but cast it into a somewhat more sophisticated and quantitative form using the advances of the half century since Bayes's work. His result, now known as Laplace's rule of succession, is based on the same likelihood function that Bayes used, where there is a total of T trials, of which N have some particular property that occurs with probability s. We saw that the posterior probability is peaked around the fraction N/T, but Laplace used the full shape of the posterior probability function to work out a better prediction for the specific question of what we should expect to happen in the *next* trial. In this case, the probability is given by the slightly different formula $(N+1)/(T+2)$.

If we have seen a million tides and no days without tides (which we could express in mathematical terms as $N = 1,000,000$ and $T = 1,000,000$), we should therefore expect tomorrow's tide to occur with probability 1,000,001/1,000,002 = 0.999999. Or perhaps it's easier to give the opposite answer: we should expect to *not* see a high tide with probability 1 − 0.999999 = 0.0001%.

Of course, this model is purely probabilistic; it does not consider the physical explanation of the tides, already understood since Newton. Given Newton's law of gravitation, the tides are completely certain as long as Earth and the moon remain in their orbits.

Even without Laplace's mathematics, Price was able to make several points here. First, we cannot be completely sure of the natural laws that govern the working of the world. Hume contended that, having never observed contradictions to the laws of physics, we should expect never to observe them in the future. But this is false at least as a matter of mathematics. Even a million confirmations of some phenomenon do not guarantee that it will happen in the future. Perhaps ironically, this is itself a mathematically precise restatement of Hume's own problem of induction—no number of observations can ever logically prove a general statement about the world.

Second, Hume had contended that we could be confident in ignoring relatively small numbers of reports of miracles, contradictions to those natural laws. But Price argued that miracles have been seen with some regularity and by no small number of observers, enough that—by Bayes's mathematics— their prior unlikelihood could be overwhelmed and leave a significant posterior probability.

From our point of view, we can see that Hume and Price were arguing past each other—they were using different models as their "background information" in Bayes' theorem. Hume was assuming the correctness of the natural laws, assigning a vanishing prior to the possibility of their contradiction— assuming the correctness of induction despite his worries. Whereas Price (possibly along with Bayes himself) was allowing their contradiction despite Newton's explication of a physical mechanism behind those laws. It seems also that they were choosing those models—their overall prior—based on their preferences for how their inferences would turn out.

Perhaps William Blake would have approved of this most radically subjective view of Bayes's probabilistic legacy: we are free to assign probabilities as we see fit. Probability theory tells us how to weave our probabilities together so that they agree with our data and are consistent with each other, but they are still conditional. Blake is free to condition upon his mythical and mystical cosmology, although I doubt he would have been concerned about the mathematical consistency of his beliefs. Hume can be confident of the constancy of natural laws, battling with Price and Bayes over their model, which gives credence to reports of miracles through the ages, despite leading to vastly different conclusions from the same data.

FIVE

Models and Measurements

By the nineteenth century, scientists were starting to get a good picture of the makeup of the objects in the sky. They had picked apart the light from the sun by passing it through a prism and carefully observed the resulting rainbow. They saw that this pattern—what we would now call the spectrum of the sun—wasn't an uninterrupted rainbow from red through violet. Rather, there were sharp, dark lines in the spectrum.

They had learned that certain gases could emit or absorb very specific colors (corresponding to specific wavelengths) of light. This, in turn, told them that the sun, or at least its outer layers, "is a miasma of incandescent plasma,"[1] with the colors of the lines telling them its constituents—lots of hydrogen and helium (itself originally discovered as a line in the solar spectrum, hence the name deriving from the Greek ἥλιος: "helios," or sun), along with smaller amounts of other gases—and the shape of the overall spectrum telling them its temperature, about 6,000 degrees Celsius or Kelvin, or 10,000 degrees Fahrenheit. What they couldn't understand is why the sun shone to begin with.

The German physicist Hermann von Helmholtz had argued in the 1850s that the sun could be powered by the slow conversion of its gravitational energy into heat and light by some unknown mechanism; he did not give any more detail for his model. The British Lord Kelvin went further. Using Newton's laws, we could measure the sun's mass. And since we knew the sun's luminosity—that is, how much energy it gives off each second—we can ask how long the sun could continue shining at that rate, taking advantage of all that gravitational energy, defined by its known mass. This gave a prediction for the age of the sun of, roughly, 20 million years, which Kelvin first published in 1862. Even then, he realized that this was something of a back-of-the-envelope calculation with simplifications that he knew were unlikely to be correct. It assumed, for example, that the density of the sun was constant, but it is natural to think that its interior would be more highly compressed than the outer reaches. This complication, he realized, might increase the age of the sun significantly, to more like 100 million years.

But even this larger value was a problem. As Darwin's theory of evolution, first explicated in *The Origin of Species* in 1859, gained adherents over the latter half of the nineteenth century, biologists realized that billions of years were necessary for the evolution of life on Earth. Darwin's work came alongside related advances by geologists in understanding the timescales for processes on the surface of Earth—the buildup of sediment in ocean beds, later seen as layers of rock; the converse decay by weathering of geographical features on the surface—also leading scientists to increase the age of Earth to the billions of years. And the sun, our source of heat and light, must be older than Earth. Kelvin, trying to justify his original calculation for the sun, also offered his own arguments that Earth might be somewhat younger. Darwin acknowledged the discrepancy in later editions of *The Origin of Species*.

But Kelvin was wrong about the age of the sun, and wrong about the age of Earth, because his model was wrong. No one then knew that there were processes more efficient than his and Helmholtz's idea of converting gravitational energy into heat and light in so efficient a way that this conversion could sustain the sun at its current brightness for hundreds or thousands of times longer than gravity alone. This conversion is, we now know, nuclear fusion, first offered as an idea in 1920 by Sir Arthur Eddington, whom we met earlier when he helped confirm Einstein's theory of general relativity by observing the bending of light rays by the sun.

Over the course of the twentieth century Eddington's idea was refined into a quantitative model of the sun, especially by Hans Bethe and collaborators in the 1930s, who, with the mathematical formulae of quantum mechanics and measurements of nuclear reactions that had become available, showed the various ways that four protons (the nucleus at the center of a hydrogen atom) could fuse into a helium nucleus, converting two of those protons into neutrons in the process. The helium nucleus is slightly lighter than the four incoming protons, and that excess mass is converted into vast amounts of energy via Einstein's formula $E=mc^2$. (These same processes are hoped to be harnessed in the laboratory to realize the dream of power generation by nuclear fusion.)

Kelvin had made the best model he could with the tools available to him at the time.[2] It was a simple and elegant model, built out of plausible physics, but it eventually contradicted the data. The model itself was exemplary: it answered a question about the world—why does the sun shine? It gave quantitative predictions that linked one observable feature of the world—the brightness of the sun—to another—the sun's age. And it was eventually superseded by a better model.

A Model Is a Story About the World

The best scientific models, like Kelvin's at first and, later, Eddington's, don't just offer a description of some phenomena. They are not just catalogs of facts but explanations that help us understand the world. They draw connections between things that might not otherwise be related, such as the brightness of the sun and the sun's age.

Darwin's theory of evolution is a scientific model that explains the variation of life on Earth. Einstein's theory of general relativity is a scientific model that explains the mutual attraction of bodies due to gravity. The theory of the Big Bang, which built upon theories of general relativity and quantum mechanics, is a scientific model for the origin and history of the Universe on the largest scales. There are scientific models for the efficacy of medications, for the formation of stars and planets, for how the liver removes toxins from the blood.

These are well-proven theories, trusted and used by scientists. But still, all these explanations are, to an extent, tentative. They try to make predictions that can be tested—experiments or measurements that we can perform to check the accuracy of the model. So they can, like Kelvin's theory of the sun, be superseded by a better theory.

Sometimes, a scientific model goes by the name of "hypothesis" or even "theory." You probably learned that scientists start with a hypothesis, and if it survives the tests of experimentation, it becomes a theory. The progress of science, like most things, is not so straightforward.

Despite what you may have learned in school, there aren't marked differences between hypotheses and theories, these different names for models. In common parlance a hypothesis may be slightly more provisional than a theory, but even a well-understood and supposedly proven theory is still subject to scientific tests and experiments. The goal, of a model or a hypothesis or a theory, is to describe the world, or some small part of the world. The process of science is, mostly, the comparison of these descriptions against the (real) world out there and coming up with new descriptions when the previous ones start to fail. We test our models, we refine them in the light of those new data, and we test them again. Sometimes, rarely, we have to throw out the whole model and start again.

Newton's theory of gravitation has withstood those tests. It is an "inverse square law," which means that the force between two objects depends on the "inverse square" of the distance between them (usually written $1/r^2$), as well as the product of the two objects' masses. The amazing thing about the law is that the masses of the objects and the distance between them are the only properties of the system that matter. But the force does depend on one more thing, a property of the Universe that gives the strength of the gravitational force. Nowadays, it's called the gravitational constant or, giving credit where it is due, Newton's constant, usually called G. Newton didn't really have enough information to measure it. He knew the distances between some objects in the solar system and something about the forces through his second law, which links those forces to the motion of the masses. But he would still need to know the masses of all the objects involved—the planets and the sun. And these are not easy to measure from afar, and even the mass of Earth itself was not known in Newton's time. The gravitational constant was not accurately measured until more than a century after Newton died, in the amazing "Cavendish experiment" of 1797–1798, which measured the very weak gravitational attraction between lead spheres of known masses and compared that force to that of the gravitational pull of Earth itself—a factor of roughly a billion times larger—giving the weight of the spheres. The ratio of those forces allowed Cavendish to work out the weight of Earth for the first time, effectively measuring the gravitational constant, although in fact the modern form of Newton's law of gravitation wasn't written down until the late nineteenth century.

Newton's law of gravitation depends on the masses of the objects involved and the number represented by G. Einstein's law of gravitation depends on both of those—the masses of the objects along with the number represented by G—in addition to the speed of light. By the time Einstein was working, the value of Newton's constant was already known, along with the speed of light. They both appear in the theories of relativity, but there are no new "Universal" numbers on which they depend.

This fact is kind of miraculous: Einstein's then new, and not yet superseded, theories subsumed gravity, electricity, and magnetism, and the motion of objects at low speeds and high, all in a way that doesn't require measuring anything new about the world. They are fully specified—sharp—theories.

Even so, these theories made new predictions, making possible Eddington's insights about the generation of energy within the sun, following on his leadership of the eclipse expedition to check Einstein's prediction of the bending of light.

Einstein's theories, with many predictions to check but no new numbers to measure, are, like their inventor/discoverer, the exception, however.[3] Most new theories have numbers that you can measure, which are then needed to fully specify the theory.

These new things to measure are sometimes called the "free parameters" of a theory. "Parameter" is just the technical term for a number that the theory needs in order to make its predictions, and "free" describes the idea that we don't know the value of that number. The theory tells us that we need to know this parameter's value and may also give us some clues for how we can measure it. Newton's gravitational constant G was a free parameter in his theory, measured since then. The masses of particles like electrons and quarks are some of the many free parameters in theories of particle physics.

A theory with free parameters can still be predictive: Newton didn't know the value of G, but the form of his equations was specific enough to both reproduce known phenomena like Kepler's laws and make further important predictions about the motions of planets and other objects. The standard model of particle physics predicted the existence, but not the mass, of the Higgs boson, proposed in the 1960s but not discovered until 2012.

The Speed of Light

Consider the idea that light doesn't travel infinitely quickly but that it has a finite velocity, "the speed of light," often denoted by the letter c. In our everyday experience, we can't tell that the first flash from a light bulb doesn't get to us as soon as we flip the light switch. The speed of light may as well be infinite, and until the seventeenth century, no one had managed to measure it. Terrestrial experiments without precision machinery couldn't tell that light doesn't simply propagate instantaneously from its source. Galileo attempted to measure the speed of light in the early 1600s by enlisting helpers on distant (but of course still visible) mountains. Unfortunately, if the mountains were a mile apart, the flash from the lantern used in his experiment would take only about 5 microseconds to travel between them. This was hundreds of thousands of times faster than the accuracy of the clocks available at the time.

The finite speed of light was eventually verified not by earthbound measurements, but by comparison with astronomical predictions. We now know that light takes about 8 minutes to get from the sun to Earth. So by considering distances between bodies in the solar system, we can go from needing an accuracy of microseconds for each mile traversed to minutes over distances between planets in the solar system.

The seventeenth-century Dutch astronomer Ole Römer observed Jupiter's moon Io going behind and in front of the planet each orbit. Using observations taken over the years 1671 to 1673, he noticed a curious fact. When Earth was moving toward Jupiter, the period of the orbit (the time it takes to go around the planet once) seemed to be shorter than when Earth was moving away from Jupiter.

The difference was only about 32 seconds compared to a total period of about 42 hours. The period was 16 seconds shorter when Jupiter was moving toward Earth, and 16 seconds longer when it was moving away.

Even this total 32-second difference would have been impossible to accurately observe with the clocks of the day. However, it wasn't necessary to measure the length of a single orbit. Instead, Römer could compare the total length of time of many orbits with the planets moving toward versus away from each other—which added up to more than 15 minutes over 30 orbits.

Where did these differences come from? Römer measured the period of Io's orbit by observing the moon being eclipsed by Jupiter, once each orbit. If Earth is moving toward Jupiter (and hence also Io), the light that leaves Io at the times of a series of eclipses arrive at Earth a bit closer together in time than they leave. Conversely, when the planets are moving apart, the signals arrive here just slightly farther apart than they leave. To get a feel for this, you can try an experiment (in your imagination or in the real world): imagine walking or running toward a wall, throwing a ball at the wall each second. Because you are moving toward the wall, the balls will hit the wall somewhat less than one second apart. In fact, the change just depends on the ratio of your walking speed to your throwing speed.

This relationship may be familiar to you from descriptions of the Doppler effect, where the frequency of a sound (or of light) increases for an object moving toward you, like a vehicle with a blaring siren, and decreases when moving away. For Römer's studies, the "frequency" was, literarily, how frequently Io reaches the same position in its orbit, when it is hidden in Jupiter's shadow.[4]

For this calculation, diagrammed in Figure 10, Römer was effectively measuring the ratio between the speed of Earth's orbit around the sun (around 30 kilometers per second) to the speed of light (around 30,000 kilometers per second), giving about 1 part in 1,000—just the 16-second change over the 42-hour period he measured. In fact, Römer, in consultation with his fellow astronomer Christian Huygens, got the actual value of the speed of light wrong by about 26 percent because he translated that ratio incorrectly to the distance between Earth and the sun. He reported that the time for light to cross the circle of Earth's orbit around the sun—which we know

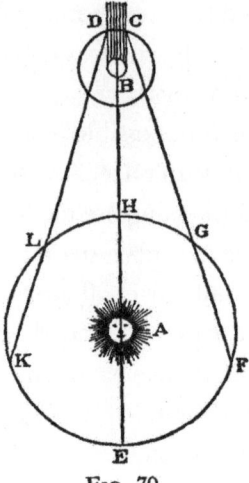

FIG. 70.

Figure 10. Römer's 1676 illustration of his method for determining the speed of light. The Earth is moving counterclockwise around the sun (A), moving toward Jupiter (B) from F to G, and moving away from Jupiter from L to K. Römer measured the change in the apparent orbital time of Io during those two periods by measuring the time of its eclipses at C and D. (Wikimedia Commons.)

to be about 183 million miles—would take about 22 minutes. In fact the actual value is about 17 minutes, and hence the speed of light as calculated by Römer and Huygens was off by about 30 percent.

By modern standards, this was not much of a measurement of the speed of light. It was wrong by a large amount, the authors didn't bother to quote any error bars, and in fact Römer never published a complete account of his methods. But after hundreds of years of terrestrial attempts, Römer was the first to realize that he could use the vast distances available to him through astronomical observations over which the speed of light could have a measurable effect with the technology available in the seventeenth century.

To get that far, Römer needed a comprehensive model including Kepler's laws of planetary motion and the further observation by Galileo that planets themselves could be surrounded by one or many moons, orbiting around them just as the planets orbit the sun. This, in turn, implied that the actual orbital period of Jupiter's moons was not itself changing. He needed to trust that his clocks were more accurate than a few minutes so he could measure the apparent change in the orbital period over almost two months. Most importantly, to be able to measure the speed of light, he had to assume that there was a such a thing to measure at all: that light didn't travel infinitely fast and that the speed of light was at least roughly the same on Earth as in the sky. This followed somewhat naturally from Galileo, again, and his no-longer-startling statement that the heavens were not so different from Earth. Newton would take this further, just a few years after Römer, of course, and posit truly universal physical laws, the naturalistic ur-model underpinning all modern science. Only with his *Principia Mathematica* published in 1687 was the physical basis for these laws first understood in terms of gravitation, the same force responsible for humans' attraction to the Earth as the orbits of planets to their moons.

Newton himself mentioned Römer's measurement of the speed of light in one of his later works, *Opticks*, in 1704, where he used another of Römer's observations to make further refinement to the theory: the shadows of Io on Jupiter were the same colors when moving toward or away from Earth. Since he had already shown that sunlight was a mixture of the different colors of the rainbow, this implied that the speed of light didn't depend on the color of the light.

Lots of theories have such free parameters to measure, like the speed of light. The free parameters of a theory give scientists the opportunity for a direct and quantitative application of Bayes' theorem. The standard model of particle physics explicitly stipulates some of the characteristics of the seventeen different kinds of particles that exist in the Universe—some, but not all.[5] We don't know the particles' masses, except by measurement, and we don't know the strength of the interactions between the different kinds of particles. That is a lot of numbers to measure, about 22 altogether. In my field of cosmology, the Big Bang theory tells us that the Universe started out hot and dense and has been expanding under the influence of gravity and pressure ever since. But the theory doesn't specify how long the Universe has been expanding, how fast it is expanding today, or the kind and amount of material in the Universe. The numbers are the "cosmological parameters," and their measurement has been one of the goals of cosmology since the Big Bang model was introduced in the early twentieth century.

These are the kinds of measurements amenable to Bayes' theorem: comparing the outputs of our telescopes or particle accelerators to the predictions of our models. Sometimes, the measurements process maps in a simple way to the parameters of the theory and, like measuring my daughter's height, I can read off the value of the parameter from a somewhat higher-tech version of a tape measure. I still need a likelihood function to account for the errors in the process, usually more complicated than squirmy kids and wonky walls, but often with the same bell-shaped curve. Sometimes, however, it's more complicated: the mass of a particle may predict the width of a bump on a chart via a convoluted formula. Or several parameters together might predict a complicated curve (this is often the case in cosmology), and going back from the data to the parameters might require complicated programs running on supercomputers for millions of hours.

But if our model is precise enough to make numerical predictions of our observations, we can apply Bayes' theorem to learn about the theory from our data.

To be really useful, such a model must both *simplify* and *generalize*. It must be more than a list of facts. If a model just catalogues the properties of some system in complete detail, without any attempt to explain, or link them together, you can't learn anything from it. Newton's law of gravitation doesn't tell us, separately, that objects fall to the ground with an acceleration of 9.8 meters per second, that the moon orbits Earth once a month at a distance of 400,000 kilometers, that Earth orbits the sun once a year at 150 million kilometers, that Mars orbits the sun every 687 days at 200 million kilometers. Instead, it gives a single simple and general formula from which all these numbers can be calculated.

Any realistic application of Newton's law is only approximately true, for many reasons. When applied in its simplest form the law requires several assumptions. First, that the objects involved are "point-like"; that is, they contain all their mass in a very small amount of space. Second, we usually assume that for any single object (such as a planet) orbiting any other single object (the sun), we can ignore the influence of any and all other objects (the other planets, their moon or moons, all the gas and small rocks in the solar system). Finally, the calculation assumes that, granting these simplifications, Newton's laws are correct—and yet we do know this to be false since Einstein's theory of gravity supersedes that of Newton and lets us calculate corrections to Newton's formula (which are small in most circumstances that we encounter).

Still, Isaac Newton's laws might, collectively, be the most important scientific model ever written down. In addition to his law of gravitation, they are accompanied by his more general three laws of motion, a package that, together, define what was in some ways the first quantitative scientific model.

This logic might all seem circular, but one of the geniuses of Newton was to realize how simple these rules made the mathematical description of all kinds of situations, from balls rolling down hills and apples falling from trees or balls falling from Pisan towers, to the moon orbiting Earth and Earth orbiting the sun. Newton's laws also help us figure out exactly how forces act and to make specific predictions that no other theory had managed before.

Newton, backed up by Galileo's observations both of falling objects near Earth and of the planets and their moons, and Kepler's descriptions of the planets' motions, realized he had a truly universal model, with the same laws as applicable to apples and balls as to planets and moons. The ability to abstract away the messy and less important details and see through to the underlying principles governing the behavior of a system is at the core of the search for scientific models.

Unreasonably Effective?

This simplification allowed Newton to see that his laws took a mathematical form. Since then, we have seen that many scientific laws are inherently mathematical—expressed as equations that describe the relationship between some quantities.

Newton, who had to invent the mathematical discipline of calculus (and thereby much of modern mathematics itself), largely eschewed equations in his work. Instead, he used a combination of words and diagrams to get most of his points across, organizing them in the form of axioms and proofs. This made his work the culmination of a tradition going back at least to the ancient Greeks, such as Euclid and Aristotle. This also makes his work much harder for the modern reader to understand (even in translation from the original Latin!).[6]

Almost all the equations you ever learn in a science class are models, or parts of models. (These equations are separate from the ones you might learn in math class, which are mostly mathematical truths, like 1+1 = 2, devoid of any physical content.)

The most well-known scientific laws and theories try to describe the mechanism that underlies some sort of phenomenon in the world. As Galileo wrote in *The Assayer* (1623), the "grand book of the Universe . . . is written in the language of mathematics."

But why does the world seem to be so well described by models given by mathematical formulae? This was a question that has worried, or at least interested, scientists and philosophers of science at least since Galileo and Newton themselves and was well expressed by Eugene Wigner in the late 1950s in a talk and essay entitled "The Unreasonable Effectiveness of Mathematics in the Natural Sciences."[7]

One way of looking at mathematics is as the manipulation of symbols on paper or clay tablets, or papyrus, or computer screens. When we first learn the more difficult and abstruse aspects of arithmetic in school—long division, say, or square roots—that is certainly all mathematics seems like: an algorithm for taking some numbers and producing some other numbers. This seems even more true if and when we learn more advanced mathematics, like algebra, or set theory. But this abstract and still very human activity was also (and primarily) invented to describe certain things that we see in the world, from counting to accountancy, not to mention physics and the pricing of financial derivatives.

We've been doing this manipulation of symbols for a long time. Over 3,500 years ago, the Babylonians had discovered some of the essentials of what we now call algebra and geometry to assist with their astronomical calculations, as well as more down-to-earth matters such as compound interest. A bit later, the ancient Greeks were concerned with some of the same things, although the mathematical-mystical Pythagorean sect were aghast at the realization, already understood by the Babylonians, that there were numbers, such as the length of the diagonal of a square, that could not be described as a ratio of whole numbers and so rejected the algebraic tools that would have required confronting this directly. Algebra itself comes from the mediaeval Arabic world—the word "algebra" is a transliteration of "al-jabr," from part of the title of a ninth-century book on the subject by the Persian al-Khwarizmi.

Especially since the time of Newton, we have had a wider set of tools including algebra and calculus and more modern creations such as group theory to help us describe the real world. How is it that these symbol-manipulation rules—algorithms—manage to do this so well that most of our best scientific theories take the form of mathematical laws?

Wigner starts his essay by noting that the mathematical symbol π appears in the formula for the bell-shaped Gaussian curve that we discussed earlier. Someone without mathematical training may very well ask why should the ratio of the circumference of a circle to its diameter (namely, π) have anything to do with, say, the description of population trends? He wrote:

> Mathematical concepts turn up in entirely unexpected connections. Moreover, they often permit an unexpectedly close and accurate description of the phenomena in these connections.

He wonders whether our mathematical descriptions are just that, descriptions of the limited phenomena we have observed, and whether different observations and observers might come up with different laws.

> It is difficult to avoid the impression that a miracle confronts us here, quite comparable in its striking nature to the miracle that the human mind can string a thousand arguments together without getting itself into contradictions, or to the two miracles of the existence of laws of nature and of the human mind's capacity to divine them.

He concludes without answering his own question:

> The miracle of the appropriateness of the language of mathematics for the formulation of the laws of physics is a wonderful gift which we neither understand nor deserve. We should be grateful for it and hope that it will remain valid in future research and that it will extend, for better or for worse, to our pleasure, even though perhaps also to our bafflement, to wide branches of learning.

In the 70-plus years since Wigner's essay, his discussion has been taken up by other physicists, mathematicians, and philosophers of science. To me, it does not seem like much of a mystery. We have chosen to describe the world with numbers: times, lengths, speeds, weights, volumes, and the like. Once we do that, our idea of *naturalism*—that the world is governed by rules that we can understand—leads to the realization that there must be relations between these different numbers. That is, there must be mathematical laws that govern them.[8]

(So if Wigner has alerted us to a mystery, there is perhaps an even larger one: Why is the world intelligible at all? Our naturalist model might be a question-begging reply to this, but the alternative, an unintelligible world, would give us nothing to understand and nothing to govern our actions and decisions.)

Of course, even granted that larger mystery, we can still ask why the laws take the forms that they do and why they involve the particular mathematical structures that they do.

Mathematics enters all sorts of theories throughout science. Cicadas are a kind of large insect with an especially loud mating serenade, "sung" by the males to attract the females of the species. Typically, cicadas are dormant for a long time, emerge for a few months to mate, lay their eggs underground, and promptly die. Although most emerge in one- or two-year cycles, in North America, there are species, known by the horror-movie term "broods," whose underground dormancy lasts for 13 or 17 years.

At first, 13 and 17 seem curious, and large, numbers. They are *prime numbers*, numbers that can't be evenly divided into other whole numbers. Biologists have realized that this puts these species of cicadas at an advantage: most other insect species have much smaller periods of dormancy between their active years—and many of the cicadas' natural predators are such other insects. The species of cicadas with long and prime periods of dormancy give them a way to avoid those predators and in particular a way to avoid those predators becoming dependent upon the cicadas.

Imagine a predator species with a 3-year dormancy period. Starting from, say, the year 2000, they might be active in 2003, 2006, 2009, 2012, and so on. If a species of cicada with a 13-year dormancy period was active in 2000, the members of this species would not meet their 3-year competitors again until 2039 (39, or 13 × 3, years later). And a species with a 17-year dormancy period that was active in 2000 would meet them in 2051. With such long times between the species' meeting, the predator would not be able to rely on these species of cicada, and so would likely evolve away from any reliance on them as prey.

But this evolutionary result is just a consequence of a mathematical fact about the divisibility of prime numbers. The theory of evolution doesn't have any "knowledge" of mathematical number theory, and the biology of these species certainly doesn't directly encode any of these mathematical theorems. Darwin's theory tells us that this elegant solution was arrived at randomly, by trying out lots of possible periods of dormancy and finding the ones that happened to work better for the species' survival. And in this case "work better" was exactly a consequence of mathematics and its reasonable (or unreasonable) effectiveness.[9]

The Laws of Light

Let's return to our story of light. After Römer's first measurement of the speed of light and Newton's first explication of general laws in the form of equations, physics rapidly became a mathematical discipline—the goal of its practitioners was to discover new formulae and apply them to a wider range of phenomena.

By the middle of the nineteenth century, physicists had discovered a set of such laws that governed electricity and magnetism. At first, these were just descriptions—models—of some curious facts about the world, such as how magnets have north and south poles, or how static electricity can literally make your hair stand on end. These were pretty exciting in their own right and had begun to have important technological ramifications, spawning the telegraph and the electric motor. Both of these took advantage of the fact that electricity and magnetism are linked.

Gauss (already known to us through his bell-shaped curve) had discovered a law that describes how individual particles with electric charge create something that we call an electric field, and a related equation that says that magnetic fields are different—all magnets have *two* poles. The English Michael Faraday had described a phenomenon called induction (this is, confusingly, the same word we have been using for making inferences from limited data, but it means something different here). His equation says that a moving magnet creates an electric field. Conversely, the French André-Marie Ampère had discovered that an electric current through a wire will, in turn, create a magnetic field. Electricity and magnetism were clearly linked, but Gauss, Faraday, and Ampère, along with other scientists and inventors like Ørsted, Henry, Lenz, Arago, Jedlik, and Morse, had only succeeded in cataloguing a bestiary of different phenomena that were somehow related.

It took James Clerk Maxwell, a Scottish physicist, to realize that he needed to modify the equations approximately describing these disparate electrical and magnetic phenomena to make them completely self-consistent. Ampère had shown that an electric current creates a magnetic field, but Maxwell showed that you can also create a magnetic field with a changing electric field, exactly mirroring Faraday's observation that electric fields are created from changing magnetic fields.

Assembling these results, Maxwell showed that all of electricity and magnetism could be described by a few short equations that link them together. His equations described what we now call the electromagnetic field: a set of quantities defined in space and in time that describe the motion of particles acting under the action of electricity and magnetism, just as the gravitational field describes the motion of particles under the influence of gravity.

Looking at his equations, Maxwell had a realization: his electromagnetic field has all the properties of *waves*. Plugging in the actual measured numbers that appeared in his equations, he saw that the waves all moved at a very high speed, a speed that was the same as the speed of light, which by that time had been well-measured to be about 30,000 kilometers per second, and indeed behaved like a wave in many circumstances. Maxwell then made a bold prediction: the wave that his theories predicted didn't just move at the same speed as light—it *was* light. His model was able to unify three disparate phenomena: electricity, magnetism, and light.

We now know that Maxwell's great leap was correct. Maxwell's laws were, and still are, a nearly perfect description of these phenomena, numerically precise and checkable with careful experiments. More practically, they give the necessary background for all the technological advances that have come in their wake, from the telegraph, telephone, and lightbulb of the nineteenth century, to radio, television, and, eventually, computers and wireless communication between them.

But Maxwell's laws are just equations. Do they tell us what is happening when a light shines? They seem to tell us that light is a wave, but what is "waving"? Sound is a wave, and a wave represents variations in the pressure and density of air (or water, or whatever else the sound is propagating through), which in turn vibrate your eardrums, where they can be detected and turned into electrical impulses in your brain. It had long been surmised that the Universe is suffused with something called the "luminiferous aether" through which light propagated in much the same way as sound through air. But all attempts to detect it failed, leading eventually to Michelson and Morley's 1887 experiment in which they tried to detect the variation in the speed of light due to Earth's motion through this aether. Instead, they found that the speed of light was exactly the same in all directions, no matter the speed of the emitter or the detector.

But in fact Maxwell's equations already encoded this. When he reduced them down to a wave equation for the electromagnetic field, propagating at exactly the speed of light, there was no room for the presence of an aether. More precisely, there was no room for the speed of light to depend on the speed of the emitter or the observer. Taken at face value, Maxwell's equations implied that the speed of light should be the same for all observers.

But this face value doesn't seem to make sense: in all our experience, if an object and I are moving toward one another, then the object would seem to be moving at a different, faster, speed relative to me. Just as the relative speed of two cars moving toward one another each at 60 mph is 120 mph, the relative speed of two spaceships moving toward one another at the speed of light would seem to be twice the speed of light.

But Maxwell, and Michelson and Morley, implied that this shouldn't be true. Einstein, like Galileo and Newton, realized that we had to be careful extrapolating from our everyday experience: none of us had ever moved any-where close to the speed of light relative to anything else; only beams of light themselves had moved at this speed.

So Einstein, also following Newton and the ancients, started with a set of simple axioms, simple facts about the world that don't need mathematics for their explication. Einstein's theory of relativity follows entirely from a simple set of rules that had been recognized by numerous physicists around the turn of the twentieth century:

1. The speed of light is the same to all observers.
2. Physical laws have the same mathematical form in all inertial reference frames.

For the second rule, I need to define "inertial reference frames." A reference frame is just a set of points that are not moving relative to each other (and so each has the same velocity compared to any other points), and an inertial reference frame has the further restriction that the points together are also not accelerating: their velocities are not changing with time.

From these principles (and some additional assumptions about how they apply when the space they are moving through isn't otherwise empty), Einstein derived his special theory of relativity, recounted in his 1905 "On the Electrodynamics of Moving Bodies," a title that shows his goal of reconciling Maxwell's equations with high-speed motion.[10] This theory had the amazing and still counterintuitive results of length contraction and time dilation: observed from the outside, moving objects seem to contract as they speed up, and their clocks tick at a slower rate.

Amazingly, the postulates led to the most famous repercussion of relativity, one that doesn't seem to have anything to do with moving bodies or electrodynamics, as noted by the equation $E=mc^2$, the statement that the energy of any object is proportional to its mass, multiplied by a universal constant, namely the speed of light squared, independent of anything else about the object.

These formulae show our mathematical naturalism at work: our simple (and in this case verbal, rather than mathematical) axioms are revealed to result in mathematical descriptions far from their starting points.

The Evolution Algorithm

After all this discussion of precise mathematical models, we might think that a scientific theory has to give complete and detailed answers to all the questions that could be posed in whatever domain it could be applied. But some scientific theories are, rather, qualitative or descriptive, and that can make them more explanatory. Darwin's theory of evolution is such a model, explaining the diversity of life on Earth. In modern terms, it is an *algorithm* that describes the necessary ingredients for organisms to become better adapted to their environments over many generations. It relies on the existence of a simple set of ingredients: when organisms reproduce, they may pass along random changes to their offspring that can continue into future descendants.[11] Because those changes may affect how well-suited the offspring is to its environment, lucky organisms with changes that make them more fit will prosper and reproduce in greater numbers than their siblings without those changes.

Because those incremental changes are random, evolution cannot tell us what traits the offspring of a specific living creature will have, and so the theory doesn't make predictions in fine-grained detail. But it does tell us that organisms give birth to offspring that will sometimes be randomly different from their parents and that some small fraction of those changes will make the organisms a slightly better fit to their surroundings. These beneficial modifications—mutations—will allow organisms with them to prosper and so more readily reproduce, with the mutations passed into future generations of their descendants. In the long run, through many of these incremental changes, this process drives organisms to be better adapted to their environments.

If evolution cannot make specific predictions, how can we check it? We can begin by examining, like Darwin, the variety of life on Earth. Organisms are indeed well-adapted to their environments, and when those environments are (like the Galápagos Islands that he visited in his research) cut off from other similar locations, species split to become more fit to their new locales. We can examine the archeological and fossil records to see that the history of life on Earth seems to match the expectations of the evolutionary algorithm, with gradual change over time adapting to the changing conditions. These

changes can be global, as the early carbon-dioxide atmosphere was trans-
formed into oxygen by the presence of the first bacteria to produce oxygen as
a byproduct of photosynthesis. This killed off most of the species then on the
planet and changed the chemical basis of all subsequent life. Later global
changes include the evolution of multicellular organisms from the earlier
bacteria, the evolution of animals, the extinction of the dinosaurs, and the rise
of mammals.

Mutations are always local, a random modification to an individual organism. A single organism can then become slightly more fit to its environment than its ancestors, its descendants accumulating further changes and
eventually cleaving off a new species. Changes to the environment can be
more widespread, as in the oxygenation of the atmosphere 2.5 billion years
ago—forced by evolution itself—or the impact of the dinosaur-killing aster-
oid a mere 66 million years ago.

Present-day organisms, and the fossil record, show evidence of both kinds
of evolutionary change. Other explanations of the same facts might be possible, but the weight of evidence in favor of the evolutionary algorithm by
now requires any alternative to be little more than a just-so story, ginned up
to explain those facts that are more naturally explained by Darwin's theory.
More importantly, because evolution is just an algorithm, and because all
the necessary ingredients—random changes passed along to offspring—are
inarguably instantiated in biological systems, evolution *must* occur.

In Darwin's time, however, it was not obvious that those necessary ingredients were part of the basic biology of organisms. Rather, the theory implies
the existence of such a mechanism for all this to happen, leaving the details
unspecified. But this, too, was eventually untangled, first by the idea of the
gene as the smallest unit of inheritance, linked to a single trait (blue eyes,
say, or the ability to digest a specific kind of plant), worked out by the Austrian monk and scientist Gregor Mendel just as Darwin was publishing his
first work on the theory of evolution. With the advent of powerful micro-
scopes and other techniques for peering inside of cells, scientists realized
over the course of the nineteenth and early twentieth centuries that chromosomes, usually located in the nuclei of cells, were the carriers of this ge-
netic information. This culminated, of course, in the discovery by Crick,

Franklin, and Watson of the structure of the DNA that makes up the chromosomes, leading to modern molecular genetics.

So Darwin's successful theory was not the end of research into evolution, but really just its beginning: it gave context for much of biological science from that point on, fleshing out Darwin's ideas and allowing previously disparate facts of biology to be understood as the results of evolution over billions of years.

"All Models Are Wrong, but Some Are Useful."

The statistician George Box wrote something like that in the 1970s, but the idea of all models being wrong is inherent in the very use of the word "models," a stand-in for some elements of the real world.[12] They are "wrong" because, even in the best case, we can never be sure about how the world really works and because almost any realistic and interesting system in the world is so complicated that a detailed description of all its parts would be impossible.

Box's quip is a model, too: a model for models. And, of course, it's wrong. (But it is useful.) It's wrong because some models are correct. Eddington's model for the sun, his idea that it might be powered by nuclear fusion, is imprecise, but not wrong. And it's even useful. But it's useful only to the extent that it gives scientists ideas for creating further, more complex models. In 1920 we didn't know enough about nuclear physics to make more than a back-of-the-envelope assessment of the amount of energy available to power the sun this way. As we understood more about both the forces holding nuclei together and the way this force governs the structure and evolution of the sun and other stars, we could make more quantitative predictions and match them to our observations of the sun. Using Bayes' theorem, and other statistical tools, we could then go backward and use the models to determine things about the sun that we cannot hope to observe directly: What is the temperature at its center? When did it first collapse and light up with nuclear reactions? How will it evolve in the future?

Even today's best models for the sun are only approximations to its full structure and history. Until relatively recently, the best solar models just treated the sun like a perfectly spherical ball, mostly ignoring obvious facts about the sun such as its rotation and its mottled, spotty surface, as well as many of the details of the way energy is carried from the center of the sun to its surface over millions of years. Recent models account for more of this complicated physics, but they necessarily still involve many approximations: after all, they don't try to model every single atom and nucleus, but somehow average over something like all of the roughly duodecillion (that's 10^{39}, or 1 followed by 39 zeros) particles in the sun at once. So even the state-of-the-art models are not quite right: they still make some approximations, even if they are much more useful than Eddington's first qualitative guess 100 years ago.

But both Eddington's and today's models are useful in different ways. Eddington pointed the way to go from a qualitative to a quantitative understanding of the sun. Today's models can predict, or at least help us understand, phenomena like solar flares, giant outpourings of energy from the solar surface that can make their way through the solar system to Earth. In the worst cases, these flares can wreak havoc with terrestrial communications and electronics, and understanding their behavior is crucial for protecting against their effects.

So these quantitative models make quantitative predictions, which can also be quantitatively checked. First, the models must reproduce the sun's mass, size, and brightness, all at the same time, and when they do so can then reveal the age of the sun even when it cannot be measured directly. But they make more predictions for numbers that we can measure. Just like they do through the air, sound waves propagate through the sun. But despite all the complicated physics required to model it in detail, the overall structure of the sun is simple enough that it acts a bit like a bell or another musical instrument: the sound waves are strongest at certain frequencies and with certain patterns of vibrations on the surface of the sun—like the vibrations of a violin string amplified by the body of the instrument. We call the study of these vibrations "solar seismology," in analogy with terrestrial seismology, which takes advantage of similar—but much less clean—vibrations of the earth in the aftermath of earthquakes.

So our solar model predicts the patterns of the sun's vibrations, many thousands of numbers that can be measured and compared against the models. We can use Bayes' theorem to help check the models and use it to tweak the numbers that go into it, in turn giving us a better understanding of the sun's structure and behavior, and its effects on Earth and the rest of the solar system.

And what we learn about the sun helps us learn about other stars, ones that we cannot observe nearly as well, although the field of "stellar seismology" lets us apply similar tools to much more distant stars, starting as far back as Kelvin's gravity-powered model for the sun—wrong, but useful insofar as Eddington and others had to come up with a better idea. And Eddington led to the more precise nuclear physics of Bethe, which led eventually to today's full three-dimensional models of the sun and other stars, around which we are even starting to observe other planets evolving under their own sun's influence by comparing models to data in detail.

In her recent book, *Escape from Model Land*, statistician Erica Thompson asks that we be especially skeptical of mathematical models.[13] A mathematical model, backed up by equations that may even seem to be confirmed by data, gives precise numerical predictions. But often, that precision is unwarranted.

Newton's model of gravity was good enough to predict the behavior of the solar system with the tools available to make measurements until about the middle of the nineteenth century. In 1859, the astronomer LeVerrier discovered that Mercury's orbit did not seem to obey Newton's laws: its shape was approximately an ellipse, but that ellipse rotated around the sun by about a degree every 700 years (an effect called "the advance of the perihelion of Mercury"), in contradiction with the earlier law of Kepler, which stated that the planets' orbits should be perfect ellipses.

Most of that difference can be understood by a more careful application of Newton's version of the law, which states that we have to consider *all* the forces acting on Mercury—its orbit is determined not just by its interaction with the sun, but by all the mass in the solar system. And indeed, by careful accounting of those other masses—particularly the effect of Jupiter, the second-largest object in the solar system—astronomers could account for about 90 percent of the perihelion advance. This was a puzzle. Had they missed any large objects in the solar system, perhaps a tenth planet? Or could the great sage Newton possibly be wrong?

This was one of the riddles that Einstein realized he had solved with his new theory of gravity: it did not make exactly the same predictions as Newton's theory and in particular predicted the observed behavior of Mercury. Even before Eddington had set out to test Einstein's light-bending predictions, general relativity had already successfully solved a scientific puzzle.

But mathematical models do not always point the way to their successors quite so easily. When we try to model very complicated systems — the behavior of humans, for example — we must make many assumptions. When designing a model to predict election results, social scientists and pollsters assume that voting behavior is at least partially determined by various characteristics of voters: their age, income, location, race, and gender, for example. This is a caricature, of course, and at best works because each of these factors is a proxy for a whole set of different aspects to any individual voter's specific opinions that eventually do determine their vote.

Usually, these models give adequate results when predicting the overall result of an election, even if their assessment of the effect of any individual factor may not be great. It works because the pollsters are able to average over many individuals with many different mixes of traits (different genders, races, income levels, etc.), and one of the most important mathematical results in statistics is that, in almost all cases, averaging gives us more accurate results than individual measurements.[14] Errors from any individual measurement will start to cancel out on aggregate, although you usually need to increase the amount of data by a factor of *four* to reduce the size of your error by a factor of just *two*.

This fact has led to the realization over the last few years that we can do even better in determining voters' preferences: there are many polls out there, and we can average them to get an overall prediction from their combination. This is the tactic pursued by Nate Silver, author of *The Signal and the Noise* and the original creator of the 538 model for US elections.[15] (The number of votes in the US Electoral College is 538 and is related to the number of senators and representatives in the US Congress.) Rather than naively combine the results, these polling averages try to reverse-engineer the individual polls' numbers to appropriately weight their results in combination, so that those polls with smaller errors count more toward the final result.

These averages have usually proven better than most individual polls. But they do require that at least the basic assumptions be correct, or at least not too wrong. If the polls are all (or mostly) systematically wrong, then no amount of averaging will help. This error is usually called "bias," and it is no different from the way that a physics model might be consistently wrong in one direction rather than randomly.

In the run-up to the 2016 US presidential election (Hillary Clinton vs. Donald Trump), most pollsters (and political pundits) gave Clinton a very high chance of winning, some as high as 99 percent even as the first returns started being announced. Some models, however, were less bullish on Clinton's chances and became less so as the election night progressed. That is because these better models—notably the 538 model—accounted for the fact that the errors or bias within a single state would not be independent from one another since the pollsters would largely be drawing from the same pools of voters.[16]

Later, some of this bias began to be understood. Most pollsters had not realized that an important factor was missing in their models: level of education. Previously, this factor was much less important in voter preferences than, say, income (with which it is somewhat strongly correlated), and it was not necessary to take it into account separately. But from around 2016, US voting patterns seemed to be strongly correlated with voter education, with the more highly educated preferring Democratic Party candidates. Prior to this time, party choice seemed to track more strongly with income than education, with higher income levels usually indicating a preference for the Republican Party. This trend had been fraying for a while, but ignoring it (and other factors) in 2016 led to inaccurate predictions—highly precise, but presented with falsely high confidence because they ignored the factor of voter education.

For science—physical science or political science—to be the self-correcting enterprise we would like it to be, we need to be able to account for the ways in which our models may fool us: all models are wrong, and only the most flexible models can account for their own deficiencies.

SIX

Physicists and Philosophers

The ingredients we've assembled in the previous chapters give us an out-line for a scientific method: come up with *models* for how things could be, compare them to the real world using *probability*, and, when necessary, revise the models based on the results. This process isn't much of an algo-rithm, and the actual history and even the present-day practice of science is messier. How do we come up with models? How do we use probability? How do we know when and how to change our models?

This description is itself a model, a model for how science should be done, and I have to admit it is wrong, like all models, in accord with Box's dictum. That is, it is not even a particularly good *description* of the work of scientists.

The basic idea that we should be comparing our ideas to what occurs in the real world has been part of the human condition as long as there have been humans and has been enforced by evolution. But explicit model-building and checking has been a part of science (or "natural philosophy" as the endeavor has sometimes been called) since antiquity, encompassing everything from Babylonian astronomy and the attempts of Aristotle and other ancient Greek philosophers to systematize their understanding of the world, through the development of agriculture, based on complicated models for the life cycles of and interactions between edible plants and ani-mals. We might even include religion as at least an attempt to model the world.

So the overall picture is, perhaps, adequate: scientists, or natural philosophers, or even priests, come up with a model to describe the world, and make at least at some attempt to compare it to the way the world really is. Even these broad strokes are not quite right: we invent our model having already seen the world and use those observations from the start to create the model.

But the role of probability has been only linked to the progress of what is now called science for the last couple of hundred years. Newton knew a little about the use of probability in gambling (and indeed corresponded with the famous diarist Samuel Pepys about dice), but probability did not figure into his scientific method. Only at the start of the nineteenth century, especially with Gauss's discussion of his eponymous bell-shaped curve, did it start to be used with any regularity in science, and only in the twentieth century did it become a standard part of scientists' toolkit for the analysis of experimental data.

Chandra

Again, we return to Eddington and his favorite models; he wasn't always right. Somewhat later in his life, Eddington came to loggerheads with a younger colleague, Subrahmanyan Chandrasekhar. (Chandrasekhar, widely known as "Chandra," spent most of his career at the University of Chicago, where I got my PhD. We students always thought that it was especially auspicious to run into Chandra—who by the time I was studying there was an emeritus professor, but still active—in the halls.)

Chandra was an Indian physicist from a well-known family—his uncle was C. V. Raman, a Nobel Prize winner. On his ocean voyage from India to England, where he was to go to Cambridge for his doctorate, Chandra made an important discovery, synthesizing recent results from quantum mechanics with the study of the structure and evolution of stars, a field that Eddington had pioneered. As a star ages it would eventually fuse all its hydrogen into helium. Once this happened, the star would lose an important contribution to its internal pressure: its own starlight, escaping from the center to the surface. That light would transfer some of its energy and momentum to the other particles in the star, pushing on them in the same way that air pushes on the surface of a balloon, providing enough pressure to keep it expanded. Once the source of starlight was extinguished, that pressure would no longer be pressing on the outer layers of the star, and they would start to collapse inward.

Eventually, astrophysicists reasoned, there would be another source of pressure that could take over, an effect of quantum mechanics whose details were just being understood at the time of Chandra's voyage, known as "degeneracy pressure." This is an effect of the "exclusion principle" discovered by Wolfgang Pauli, who showed that matter particles such as electrons and protons behaved in what seemed like a very strange way: no two identical particles can occupy the same state at the same time. "State" is a term of art in quantum physics, referring to not just a particle's position, but all the other variables needed to describe it, and "degeneracy" is another technical term for being in the same state. Indeed, this is not an esoteric phenomenon: degeneracy keeps the molecules of solids and liquids from overlapping and enables the world to be made of distinct substances and objects.

As gravity compresses a star ever further, its electrons will move ever faster, but simultaneously repel each other due to the need to stay out of the same state, increasing the pressure pushing back. Eventually, they will move at speeds approaching the speed of light, and Einstein's theory of relativity will need to be considered along with quantum mechanics.

Chandra's innovation was to combine all these ingredients: relativity, degeneracy pressure, and the evolution of stars. He realized that the so-called white dwarfs—small, dim, and incredibly dense stars whose properties were just beginning to be understood in the 1920s—would have to experience this relativistic degeneracy pressure in their centers if they were massive enough, and that this pressure would lead to an upper limit on their mass, beyond which the election degeneracy pressure would be insufficient to keep them from collapsing.

Chandra figured all this out with pencil and paper while traveling to pursue his PhD and published a series of papers in the early 1930s discussing the physics behind this discovery. Although the calculations were careful, and laborious, Chandra's result was remarkable in its simplicity. The maximum mass of such a white dwarf—now called the Chandrasekhar mass—could be expressed in a beautiful equation, using just a few of the fundamental constants that describe the Universe:

$$M_{\text{Chandra}} = \left(\frac{\hbar c}{G}\right)^{3/2} \frac{1}{m_{\text{H}}^2}.$$

Even though this equation might look daunting, each of the symbols is there for a very specific reason:

- G is Newton's constant, necessary because *gravity* is holding the white dwarf together;
- $\hbar$ or "h-bar," Planck's constant, is the fundamental constant of *quantum mechanics*;
- c is the speed of light, coming from the theory of *relativity*; and
- m_{H} is the mass of the hydrogen atom, the chief constituent of *stars*.

The full equation has other numerical constants out front, accounting for details of the structure and history of the star. Those turn out to be not far from one, and when we put everything in carefully, the Chandrasekhar mass ends up about 1.4 times the mass of the sun. And indeed we've never seen a white dwarf any more massive than that.

But that limit implies that some stars have a very disturbing future: rather than cool off over time as a white dwarf, a sufficiently massive star will not be able to hold itself up once it stops burning its nuclear fuel and creating radiation to counter the gravitational force. Gravity will win out and the star will continue to collapse, until we need to take Einstein's full theory of relativity into account. That is, it will collapse into a black hole.

And in the 1930s, this possibility was seen as untenable by many astrophysicists, most notably Sir Arthur Eddington. Eddington had participated in the creation of most of the theory of stellar evolution up to that point and, as we have already seen, was a champion of Einstein's relativity. Eddington was one of Chandra's PhD examiners, and the two became friends in the early 1930s. In 1935, Chandra's work was presented to the Royal Society in

London, along with an unexpected and lengthy response from Eddington. He could not imagine that nature would allow real black holes to form and rejected the premises and results of Chandra's theory, instead insisting — with a mistaken theoretical justification — that relativistic degeneracy itself would not occur: "I think that there should be a law of nature to prevent a star from behaving this way"[1] — this is a Bayesian prior of sorts, but a dangerous one.

In his published paper on the subject, Eddington wrote

> Using the relativistic formula, [Chandrasekhar] finds that a star of large mass will never become degenerate. . . . When its supply of subatomic energy is exhausted, the star must continue radiating energy and therefore contracting — presumably until, at a diameter of a few kilometers, its gravitation becomes strong enough to prevent the escape of radiation. This result seems to me almost a *reductio ad absurdum* of the relativistic formula. It must at least rouse suspicion as to the soundness of its foundation.
>
> I do not think that any flaw can be found in the usual mathematical derivation of the formula. But its physical foundation does not inspire confidence, since it is a combination of relativistic mechanics with non-relativistic quantum theory.
>
> In the present paper this unholy alliance is examined.[2]

Eddington's simple picture — all stars end up as white dwarfs — was intuitively appealing, but rested on mistaken calculations of the effects of degeneracy in the centers of heavier stars. Many other physicists agreed with Chandra's analysis — a few in public, many more only in private — but Eddington's fame and personality were too strong, even in the face of other published papers agreeing with Chandra's results. Without Eddington's backing, Chandra felt that he could not remain in Britain and accepted a position at the University of Chicago in 1936, remaining there for the rest of his life.

Undoubtedly Eddington's stance affected Chandra's career (although as an alumnus of Chicago, I cannot claim it was for the worse!). Moreover, it contributed to theoretical misunderstandings in both the theory of the evolution of stars and in general relativity itself, in particular the delay in detailed understanding of black holes until at least the late 1950s, with Chandra himself revisiting the subject in the 1960s and 1970s.

Chandra—who had a long and influential career—occasionally seemed embittered by the interactions with Eddington, of whom he said "despite this man's incredible physical insight, he has always operated with preconceived ideas."[3]

Eddington died in 1944, and in a speech delivered in his honor Chandra said (in Chicago, where he had settled),

> I believe that anyone who has known Eddington will agree that he
> was a man of the highest integrity and character. I do not believe
> for example, that he ever thought harshly of anyone. That was
> why it was so easy to disagree with him on scientific matters. You
> can always be certain that he would never misjudge you or think
> ill of you on that account. This cannot be said of others.[4]

In 1982, Chandra delivered lectures, on the centenary of Eddington's birth, entitled "Eddington: The Most Distinguished Astrophysicist of His Time." The lectures were collected into a book in 1983—the year Chandra was awarded the Nobel Prize for his work on white dwarfs and the structure of stars.

Models of Science

How can we understand this seemingly irrational behavior, of Eddington himself (who never recanted his position on white stars before his death in 1944) and of at least some of his contemporaries who followed his lead in disagreeing with Chandra? Or of Kelvin's continued attempts to reconcile his theory for a purely gravitational source for the sun's luminosity even in light of contradictory information about its age?

These questions remind us that science is a much messier affair than the rather austere methods of model-building and checking that I have been describing. We don't just keep a table of theories and their probabilities, calmly updated when new data arrives or when we come up with new wrinkles on our theories. Instead, we are constantly poking and prodding at whatever theories we currently have at hand, trying to understand the results of our experiments and the repercussions for those theories, sometimes picking winners and losers (for our attention, for funding, for publication in prestigious journals) often based on opinion and personality along with empirical evidence.

Philosophers have been wondering about the right way to learn about nature for hundreds of years. Until the nineteenth century, there wasn't a sharp distinction between scholars whom we would now call "scientists" and those we would label "philosophers." Indeed, much of science fell under the rubric of "natural philosophy." Because of this, it was usually those scientists themselves who thought about how science ought to be done. How do we come up with new ideas? How do we test them? How do we tell a good idea from a bad one?

So the real model for the progress of what we now call science has been determined by scientists themselves—learning by doing and under the apprenticeship of our predecessors.

One strand of ideas goes at least as far back as the Renaissance English philosopher and politician Sir Francis Bacon in the early seventeenth century. Bacon advocated testing theories via a *crucial experiment* that could support one theory against others, although, anticipating Hume's problem of induction, it couldn't prove that theory to be correct.

Most scientists would probably subscribe to something like the version of this idea that was developed by the mid-twentieth-century Austrian-British philosopher Karl Popper. His many contributions include *The Logic of Scientific Discovery*, a book that attempted to describe the difference between theories that were scientific and theories that were, well, everything else. If we are truly doing science, Popper said, we only look for theories that are falsifiable, which make novel and verifiable predictions. If the prediction fails, the theory is false. If we keep believing the theory, we are not being scientific. And if a theory can't be falsified, it isn't a scientific theory. Scientific progress only happens when a theory is falsified by new data and, we hope, superseded by a theory whose predictions can match the data—and ideally makes further new predictions. Along the way, Popper claimed to have solved Hume's problem of induction, or at least dissolved it, by allowing us to falsify inductive statements even if we can't confirm them.

Our observations are rarely so clear-cut. Eddington's result for the deflection of light by the sun wasn't a single, precise angle, but rather a number with error bars, describing the probability of a certain amount of deflection. So, at best, we could say that the Newtonian prediction was probably false. We could attach more precise numbers to those probabilities, but "probably false" is not really a deductive statement like "Socrates is a mortal."

Popper knew enough about the reality of scientific practice to understand this. This idea of falsification is presented in his book as the logical idealization of the scientific method rather than a description of scientists at their desks or telescopes or laboratories. Popper sets up this ideal because he is pursuing a second goal: he wants to solve the problem of "demarcation," to distinguish between science and nonscience.

Some theories are not falsifiable and hence, according to Popper, are not scientific. For example, Popper included Marxist economics in this category. Marx himself made specific predictions for the evolution of capitalism into monopoly and its eventual replacement by communism through revolution in the most advanced industrial nations. When these did not come true, twentieth-century Marxists were able to continually make ad hoc adjustments to the theory to accommodate any new data, the sign for Popper of an unscientific theory, never able to be refuted but only confirmed by the machinations of its adherents in the face of whatever data come along.

This *model* for the progress of science, and for the demarcation of scientific and unscientific theories (and it is a model), is beautiful, simple, . . . and doesn't really describe the more chaotic way science actually proceeds.

One hint of the problem is shown by the so-called Duhem-Quine thesis (Duhem was a late nineteenth- and early twentieth-century physicist; Quine was a wide-ranging twentieth-century philosopher and logician). They each noted that we never really a test a theory in isolation. Rather, we must accept a plethora of other statements even to start testing a theory. We need to understand the working of our instruments and the different ways that they might fail—our apparatus will usually give us only noisy estimates of the quantities we are measuring. We need to understand and believe any other theories that are related to the theory we are trying to test and the experiments we are using to test it. For example, to really trust our astronomical measurements we need to understand the propagation of light through empty space, and through air, and through the glass of our telescopes. When Eddington (yes, again) reminded us that, having accepted the possibility of gravitational lensing, we need to ensure that it is not corrupting our measurements, he is putting forward a version of this thesis. When we consider the possibility that our bell-shaped curves allow our results to be, rarely but occasionally, off from the right answer by very large amounts, we are also invoking it.

Really, Duhem and Quine remind us that we are always testing the combination of our theory and all these "auxiliary" statements, which may themselves be the product of other theorizing that we have made. We can't test ideas completely in isolation.

A somewhat different model was offered by Thomas Kuhn, who took a more sociological view of scientific progress, as practiced by communities of scientists. "Normal" science comprises research under the umbrella of a so-called paradigm that researchers take more or less for granted. Most of the time we are doing such normal science, trying to make our given theories more precise. He contrasted this with "revolutionary" science, which is when science breaks the dominant paradigm only when it can no longer accommodate the data. In the first case, scientists use the prevailing theory, filling in its gaps and working out its repercussions, assuming and usually not questioning its truth. (This is of course a simplistic description. Perhaps because many scientists have read Kuhn, we do in fact hope to discover something in contradiction with the dominant paradigm, hastening the coming of the revolution.)

The canonical example for both Popper and Kuhn was the overthrow of Newtonian mechanics by Einstein's theory of general relativity. We have already seen how Einstein's theory accommodated the already known precession of Mercury's orbit and predicted the lensing of light by the sun as observed by Eddington's eclipse expeditions. After these triumphs, Popper would say that the Newtonian theory had been falsified; Kuhn would say that normal science under the Newtonian paradigm was impossible and that it had to be overthrown by the new relativistic paradigm.

The Baconian/Popperian idea of falsification, modified to confront the critiques of Duhem and Quine, culminated in the work of Imre Lakatos, a late twentieth-century Hungarian-British philosopher. Lakatos incorporated Kuhn's insights about the progress of normal science, contending that the object of scientific research wasn't a "theory" but a "scientific research program." A scientific research program consists of a *core* set of theoretical statements complemented by a *protective belt* of auxiliary hypotheses. If we can get through the protective belt and disprove the core, we have falsified the theory. These dis-proofs reduce the scientific method to deductive logic, just like Popper said. In fact, this method is often described as "hypothetico-deductive": we make (logically sound) deductions about the truth of (falsifiable) hypotheses.

Consider Newton. The core of his theory was the set of his laws of motion, along with the law of gravity. But these are not enough to say anything at all about what we expect to see in the night sky. For that, we need a large set of auxiliary hypotheses. We need to know which of the lights in the sky are planets, we need to know that they are small compared to the distance between them, and we need to know that there are no other large, heavy objects around them to perturb their Newtonian orbits. Indeed, this latter assumption sometimes turns out to be false, and it was necessary to jettison it from the protective belt when the orbit of Uranus, discovered in 1781, failed to match observations. Instead, the overall scientific program was modified to include a new planet, with calculations by astronomers in Britain and France predicting its location, culminating in the direct discovery of the new planet, Neptune, in 1846. This ability of the Newtonian program to be modified and to make further testable statements was the mark of what Lakatos called a "progressive" program. But eventually, even the Newtonian program failed. Mercury's orbit didn't quite obey Newton's laws, but no new object could be found to be responsible for the differences. Instead, it matched the predictions of a new scientific program, that surrounding Einstein's theory of general relativity, another (still) progressive program.

Lakatos put these progressive programs in contrast to what he called "degenerating" programs of scientific research. These programs fail to agree with data, but rather than being modified in a way that made new predictions, they would either just continue to fail to agree with data or—worse—would be modified in a way that accounted for just this discrepancy but made no new predictions. These are theories that are just tweaked with ad hoc modifications to accommodate whatever is thrown at them. Lakatos, like his forbear and mentor Popper, put Freudian psychological theory into this category, along with Marxist economics.

In astrophysics, we have already seen a degenerating example in Kelvin's defense of his theory that the sun could be powered by gravitational energy. In order to defend the core of the theory, he had to adapt the auxiliary belt needed to describe the age of Earth and the structures on it, searching for ways to increase the age of the sun and decrease the age of Earth, despite what became the prevailing, and seemingly correct, evidence that they were both between 4 and 5 billion years old, rather than the few hundred million years that Kelvin predicted. The efforts of Eddington—who had understood Kelvin's error—to save the evolution of massive stars from the fate of surpassing Chandra's limiting mass might also have put the theory of stellar evolution into the degenerating category, but Chandra eventually won out, and we now see that black holes are indeed the end for many stars.

Anything Goes?

Paul Feyerabend was Imre Lakatos's good friend—and intellectual adversary. He had, perhaps, the most accurate description of the scientific method in its actual practice over the last several hundred years: "anything goes," an idea elaborated in his tellingly titled book, *Against Method*.[5] Feyerabend argued that there simply wasn't a single theory of knowledge that describes how we got to where we are and that the history of science certainly wasn't anything like the model-building-and-checking paradigm that I've elaborated, nor is it the hypothetico-deductive ideas of Popper and Lakatos.

Where Lakatos saw his philosophical program as providing an overall model for how scientists should (and sometimes do) work, Feyerabend thought that the lesson of Duhem, Quine, and Kuhn—and of the real history of science—was that all this effort was in vain: there was no unified model for scientific progress that applies to the real world. Science was a vibrant, successful, and effective human activity, but like all the things that we do, science is superficially rational with an underpinning of chaos, emotion, and politics. And so, it is impossible to reduce it to a set of practical rules.

He saw this in examples like Eddington and Kelvin (although he never discussed these cases explicitly) and even, or especially, in the transition from the ancient Earth-centered Universe through Copernicus's heliocentric view—with its circular orbits that failed to predict in detail the known facts of planetary positions on the sky—through Galileo and eventually to Kepler's more correct theory of ellipses, and Newton's eventual mathematical description.[6]

Neither Lakatos nor Feyerabend thought that an inductive procedure could work as a basis for the scientific method. Following Hume—and Popper, and Duhem, and Quine—the logical flaws of induction seemed insurmountable as a path to certainty. As Feyerabend described his own reaction:

> "Anything goes" is not a "principle" I hold—I do not think that "principles" can be used and fruitfully discussed outside the concrete research situation they are supposed to affect—but the terrified exclamation of a rationalist who takes a closer look at history.[7]

For sure, attempting to reduce the practice of science to the search for logical certainty is doomed—Hume's problem of induction is real. There is not much we can learn without models to limit the scope of our explorations and without embracing probability in place of certainty.

Feyerabend famously argued for "counterinduction," allowing us to consider theories that we know to be false but may explain some of the data at hand, or have other nice properties such as simplicity:

> The ... "counterrule" which favours hypotheses inconsistent with observations, facts and experimental results, needs no special defense, for there is not a single interesting theory that agrees with all the known facts in its domain.[8]

This description is a bit too cute: as a practical matter, some interpretations of the data are bound to be incorrect—but knowing which results are safe to ignore takes experience and knowledge, and even then, most attempts at such counterinduction are unsuccessful.

But Feyerabend's program (although he would not label it as such!) aligns with some of the probabilistic approach I have espoused: all probabilities are conditional. And hence you are free to condition upon whatever theories that you wish and consider whatever data you wish.

Once you have probability as part of your scientific method, induction and deduction become two ends of a spectrum: one retaining the best-fitting theories, the other discarding the worst-fitting ones. But the use of probability is another area on which the philosophers, and philosophically minded scientists, disagree.

One school of thought takes the Bayesian use of probability seriously: we select models by calculating their conditional probabilities given our data and the relevant background information. This is the scientific method that I have advocated in this book, based on the question "How probable is my model, given my experiments and observations?" You specify your background information and the model you wish to test (or the parameters you wish to measure), gather your data, and combine these with Bayes' theorem to assign the prior and likelihood needed to calculate the posterior. Comparing the posteriors for different models, or different parameters within a model, gives you everything you need to know—everything that there *is* to know—about your model in the light of the data.

So I measure my daughter's height and each time get a slightly different answer, not only because she has grown, but because of all the things out of my control: her squirming, the imprecision of the ruler and my eyes reading it, the slightly different bend of our wonky walls depending on exactly where she stands. If I assume this likelihood is described by the usual bell-shaped Gaussian curve, I can assign appropriate error bars and understand how close I am likely to be to her true height, when combined with any prior information I may care to include. The exact same process applies to properly scientific measurements—the distance to a star, say, with each experiment's own different sources of noise and error corresponding to its bell-shaped (or other) curve describing the errors.

The other school of thought, often called "orthodox" or "classical," more closely aligned with the prescriptions and descriptions of Popper and Lakatos, requires the scientist to calculate not the probability of the model given the data, but only that of the data given the model. It asks the question "If my model is true, how likely am I to have gathered the actual data that my experiment produced?" To answer this question, this method relies on the frequentist picture of probability that I discussed earlier. We imagine an infinite number of repetitions of my experiment or observation, assuming the truth of the model but allowing all the sources of noise and randomness to vary each time.

Now, I can start with the same measurements, and the same bell-shaped likelihood. But after I read off the height, the orthodox methodology demands that I ask the subtly different question "If the height were really some particular number, how likely would I have been to make the measurement that I did?" And in fact it's even worse than this: the full frequentist algorithm demands that you ask the question "How likely would I have been to get a result *even more discrepant* than what I measured?" If this sounds convoluted, well, yeah. Instead of fixing the data to those observed, but comparing different models in the face of your data, this interpretation of probability only lets you apply it to imagined or real repetitions of some "random process." Instead, the classical/orthodox/frequentist scientist cares mostly about testing the efficacy of her procedure for measurement and analysis.

Both the Bayesian and frequentist will often report their results as an error bar around some value. For the Bayesian, the error bar represents a *credible interval*—given our data, likelihood, and priors, it represents the region within which we are (say) 68 percent confident that the true value lies. For the frequentist, the error bar is a *confidence interval*: it is a region that contains the single, fixed, and unknown true value of the parameter in (again, say) 68 percent of possible repeated experiments—even if those experiments are never actually performed.

Despite these two very different interpretations of error bars, in many cases the two methods for constructing them will give similar or even identical results. This had led to some complacency among practical scientists, who just want to get on with their fun jobs creating models and doing experiments: Maybe it doesn't matter how I think about this part of my data analysis? Unfortunately or otherwise, for more complicated (read: "realistic") problems, the practical differences between the two views of probability can matter.

And in fact these two schools of thought have been at odds for at least a century. The Bayesians were accused of unscientific subjectivity—the requirement for assigning a prior to the parameters we wish to measure. For their part, the frequentists were charged with doing objective calculations of probabilities that we don't care about—why should the probability of getting a measurement as discrepant or worse than my actual result tell me anything at all about the value I'm trying to measure?

Both accusations have some truth, but the facile slogans ("Bayesians aren't objective!"; "frequentists don't actually measure anything useful!") don't grapple with the real-world successes and failures of both methods.

For much normal science (co-opting Kuhn's term) either method is fine: as we acquire more and more data, the error bars shrink to encompass a small range around the true value. Inductively, we can be reasonably sure that we are near the true value. Deductively, we can be reasonably sure that we cannot disprove a model within the error bar.

As a practicing scientist, I am a pretty thoroughgoing Bayesian and admit to finding it difficult to offer a straightforward description of the frequentist interpretation! But I must also admit that the probabilistic Bayesian account that I am advocating is also an idealization. Most scientists are not rigorous Bayesian probabilists, and it would be impossible to scour the scientific literature and create a web of consistent probabilities capable of rigorously measuring the parameters we care about and adjudicating among the possible theories.

So perhaps I'll have to join in Feyerabend's anarchic cry of "anything goes."

PART TWO

The Randomness of the Universe

PART FOUR

The Mathematics of the Universe

SEVEN

Entropy and Information

We need the mathematics of probability because so much of the world around us seems random: we can't predict what's going to happen and we want to assign probabilities to the different possible outcomes. But what does it mean for something to be random? The *Oxford English Dictionary* defines "random" as "made, done, or happening without method or conscious decision," but it also gives a technical definition from statistics: "governed by or involving equal chances for each item."

So we find the word "chances" here—and the presence of "chance" reminds us that we are dealing with probability. And this definition mirrors our old tool from Laplace via Keynes: the principle of indifference. We applied this principle when we wanted to assign probabilities to several possibilities, none of which we thought to be more likely than any other. In this case, we assigned equal probabilities to each possibility: 1/2 to a flipped coin landing on heads or tails; 1/6 to a rolled die landing on a particular side; 1/7 to which day of the week some arbitrary event might occur (if there was no reason to think it might be more likely to happen during the weekend or the working week). Certainly, we would think of those possibilities as being random in this sense of the word.

The statistician Richard von Mises (not to be confused with his brother, the "classically liberal" economist Ludwig von Mises) realized that a good working definition of "random" is "the complete absence of any rules that may be successfully applied to improve predictions about the next number."[1]

Note the phrase "improve predictions": von Mises is saying that not only can I not predict exactly what is going to happen, but that there is absolutely no pattern that gives *any information at all* about what is or isn't going to happen. To make money gambling (for example), I don't need to predict the exact outcome each time. I just need to improve my chances, enough so that I win more than I lose. If the outcomes are truly random by this definition, there is nothing I can do to increase my chances at all.

Throwing a single die is a good example of such a "random number generator." If the die is a uniform cube and thrown fairly, no one should be able to predict the outcome of which side it lands on—as before, we would assign the same 1/6 probability to each outcome.

But what if—as in a real game of craps—we throw two dice and care about the value of their sum? For this game by adding the face values of each die, we can get a total of 2 to 12, but the probabilities are no longer equal. Instead, they are proportional to the number of different ways I can make each value. I can only make 2 by getting a 1 on each die, but I can make 7 by several combinations: 1+6, 2+5, and 3+4 as well as 6+1, 5+2, and 4+3, which are different since the order of the dice matters.

In craps, there's a pattern: the outcome is still random, but it is no longer *uniform*. That's how the casino sets the betting odds in craps: 7 is the likeliest outcome, 2 and 11 the least likely, with the probability of the remaining numbers climbing and falling in between, as seen in Figure 11.

Imagine someone rolled a pair of dice many times and recorded the sums and also rolled a single twelve-sided die and recorded the outcome. Can I compare the lists and tell which was most likely to have been generated by the single die or the pair? (There's a subtlety here: the sum of two dice will never give 1, since the lowest possible roll is 2 (i.e., 1+1 = 2). So the presence of even a single occurrence of the number 1 on the list would prove that it came from the single twelve-sided die—a bit like Popper's falsification of the two-dice hypothesis via a single contradictory experiment. To make it more interesting, let's imagine that I ignore the 1 outcome from the twelve-sided die.)

With only a few rolls, I probably can't tell which of these sequences came from the twelve-sided die and which from the sum of two six-sided dice:

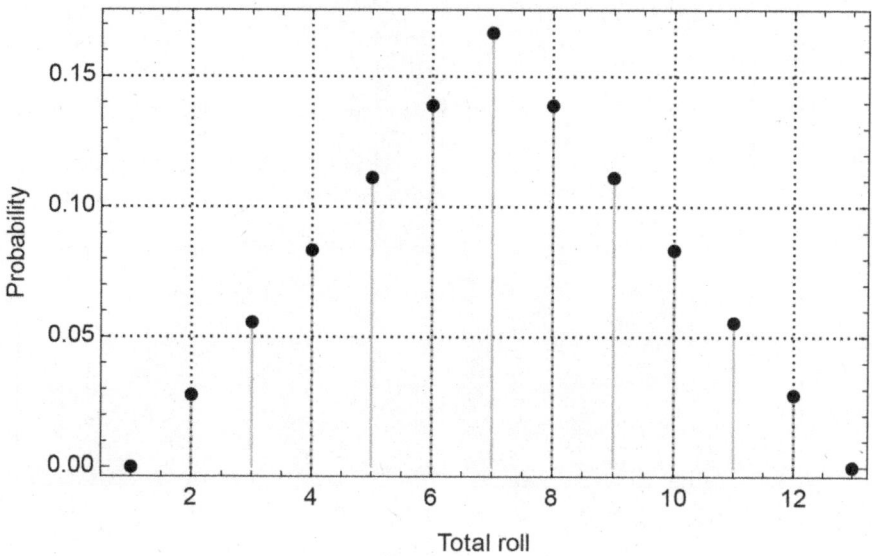

Figure 11. Probability for different outcomes for rolls of two dice. Note that there is zero probability of rolling 1 or 13.

10, 12, 6, 11, 7, 8
12, 8, 2, 9, 4, 11

But once I've got enough rolls, it becomes easy to tell which set of numbers came from the single die or the pair. I won't write out the numbers for a list of rolls for each, but instead Figure 12 shows a pair of plots for the two cases, where the height of the bar gives the number of rolls with each outcome. Now, it's easy to tell them apart: the one with the peak around 7 is much more probable to be the sum of two six-sided dice. (More probable, but not guaranteed!)

The rules, and the betting outcomes, of craps are made to account for these differences in probabilities—to the advantage of the casinos. They have chosen the betting odds for games like craps exactly so that knowing the shape of this curve (for craps, or roulette, or any of the other games played against the house) doesn't give you any advantage.

Really, we need to expand our definition of randomness somewhat. As with probability, randomness is conditional and depends on the information

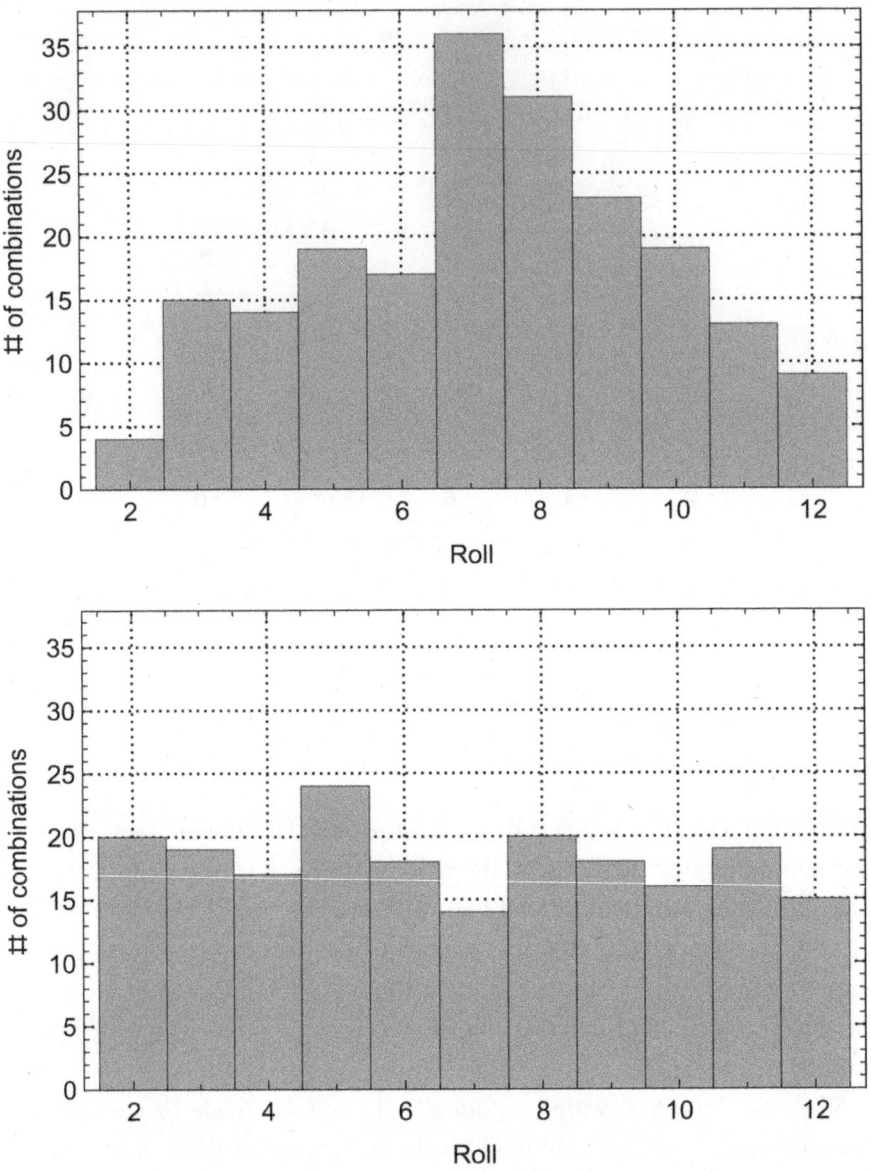

Figure 12. Chart of the number of times a given number occurs in the roll of two dice versus a single twelve-sided die. Can you tell which is which?

that we have. The sum of two fairly thrown six-sided dice is completely random, conditional upon its triangle-shaped distribution of outcomes. But it's not completely random relative to the flat or uniform distribution between rolling a 2 and a 12. If I was allowed to place bets with uniform odds that the sum of two dice would be between 2 and 12, but I knew that you were rolling two dice and summing the outcomes, I could bet on 7 and be likely to make money on average. But (as in a real game of craps), if the betting odds consider those probabilities, then I cannot game the system.

Our definition of randomness encompasses both kinds of processes: simple ones, for which the principle of indifference may hold, as well as more complicated ones where there might be a known probability distribution, but there is no information beyond that distribution that can help us make predictions.

Kinds of Randomness

Von Mises realized that any bias that would enable a gambler to game the system would be a departure from complete randomness. The outcome of a trick die that always rolls a 3 is certainly not random. But a loaded die, weighted so it comes up 3 about one-third of the time, instead of the usual one-sixth, is also biased, and so not completely random, even if we can't predict exactly which rolls will show the 3.

So there are many—infinitely many—ways to depart from randomness. How can you describe all of them? The mid-twentieth-century mathematician Per Martin-Löf realized that a true departure from randomness—a true bias—arises only when you can describe a sequence more concisely than just listing the values in it.

One such concise description could be a mathematical formula that enables you to calculate the next value in the sequence exactly. That is the simplest version of a nonrandom sequence. And since there are infinitely many mathematical functions that you can write down, there are infinitely many nonrandom sequences.

But departures from randomness don't have to be so prescriptive as to give exactly the values in the sequence. Even a formula for the probability distribution—like our dice-sum triangle—would help.

This is how the Morse code system that was developed for the telegraph worked. Morse code had to translate all the letters in the alphabet as well as the ten numerical digits into a code using only two codes, known as "dots" and "dashes." For example, you might just start with the shortest patterns at the beginning of the alphabet and assign them a string of codes, starting from A and progressing through Z and then from 0 through 9. But English words aren't just a random collection of letters! Morse code takes advantage of the fact that some letters—E and T for example—are much more common in English than other letters such as Q and J. It then ascribes shorter patterns of dots and dashes to the common letters and longer patterns to the less common ones. Most telegraph messages will end up shorter than if the code didn't take this into account.

This procedure is also how data compression works on computers. You might have downloaded a zip file, which is a format for compressing any file on your computer, taking advantage of any nonrandomness of the file. Rather than starting with the Roman alphabet, it just treats the original file as a sequence of numbers (computer bits and bytes) and analyzes it for departures from uniform randomness, just as we were able to compare the bar charts for the sum of two dice versus the single twelve-sided die. It then creates a bespoke Morse-like translation dictionary for this particular file, giving shorter codes for the most common sequences in the file and longer codes for the less common ones. This encoded version of the file—along with the dictionary itself—gets written to the disk as the compressed version of the file.

Of course, not all files are compressible: the more random the file seems to be, the harder it is to compress. As a corollary, this implies that a compressed file should appear to be random—otherwise, it could be compressed even further.

This requirement for a concise description of the departure from randomness is related to an alternative description of randomness itself, this time related to suggestions from Kolmogorov, whose axioms of probability we encountered before. According to Kolmogorov, we'll call a sequence random if we *can't* find a shorter description of it. For a random sequence, for example of numbers, the best we can do is just list the digits; for a nonrandom sequence we can always compress it down, even slightly, by a suitably clever formula or computer program.

It turns out that these definitions of randomness are more or less mathematically equivalent, whether we talk about betting on an unbiased sequence, or concise descriptions of the departures from randomness, or concise descriptions of the sequence itself.

So one way to think about our probabilistic models of the world is with random numbers. Once we have *assigned* a probability distribution, we can think of our model *as if* there was a random process spitting out numbers from that distribution each time I make a measurement.

The model for the measurement of my daughter's height, or of the distance from Earth to a star, takes the right answer and adds a random number to it, where the random number is described by one of our bell-shaped Gaussian curves, centered around zero. That's not what really happens, of course: the error around the right answer is really due to a myriad of small effects (the wonky walls, the squirmy child, and so on) adding up to something that we end up describing by the bell-shaped curve. Randomness arises when we do not have precise models for the outcomes of our experiments. Like probability, any randomness in our models reflects the information available to us.

Von Mises, Martin-Löf, and Kolmogorov have taught us how to recognize a random sequence if we see it. But how do we create seemingly random numbers? In practice, we want to use computers to be able to simulate the world, including the parts that we describe with our probabilistic models.

Computers are very good at following instructions. That's just what every computer program is: a set of instructions for the machine to follow, to the letter, every single time. That's why your word processor or spreadsheet opens the same way every time. But even spreadsheets, which are computer programs designed for working with numbers used by businesses and individuals every day, have formulae for generating "random numbers." But one of the hallmarks of a random number, according to von Mises and his successors, is that there shouldn't be a concise formula for predicting random numbers. And basically the only thing that a computer can do is follow concise formulae—computer programs.

So how can a rule-following computer make random numbers? First, they start with an arbitrary number called a *seed*. Then, they apply a known (and not necessarily very complicated) algorithm to that seed, which produces a sequence of new numbers that seem random. For example, a simple random number algorithm is

1. Pick a seed between 0 and 8.
2. Multiply the number by 4 and add 1.
3. Divide by 9 and take the remainder.
4. Go back to step 1 using the remainder as the new starting point.

So, for example, let's start with the number 3. We multiply this by 4 (for a total of 12), add 1 (for a total of 13), then divide by 9 (giving 1 with a remainder of 4). Now, according to step 4, we go back to the first step using the remainder, 4, which we multiply by 4 (for a total of 16), add 1 (for a total of 17), then divide by 9 (giving 1 with a remainder of 8). If we continue this process, the final number produced in last step each time we move through the steps will appear in this order in the following sequence: 6, 7, 2, 0, 1, 5, 3. Once we get to 3, the algorithm—which only depends on the previous random number—repeats, giving the numbers 4, 8, 6, 7, 2, 0, 1, 5, 3.

This sequence obviously isn't "really" random—I've just given you a formula for it, while the whole point of a random sequence is that it should be unpredictable. But if you are unaware of the algorithm—and if you didn't look at more than one round of nine numbers, you would not be able to predict the sequence.

And if we changed the second and third steps, we could make the sequence repeat after a much longer list so that we'd probably never even notice the repetition. For example, if we chose to multiply by 69,069 in step 2 and divide by 2^{32} in step 3 before taking the remainder, the sequence would not repeat until it worked its way through all of the more than 4 billion numbers between 1 and 2^{32}. But the algorithm still would be predictable in principle, if not in practice.

These sorts of algorithms are called *pseudorandom number generators*, and computer programs use them all the time. I used them to simulate the roll of dice mentioned earlier. If you've ever played a computer card game, its programmers almost certainly used a random number generator to make sure that the simulated deck of cards is shuffled differently each time, or that your computer-generated adversaries don't behave the same way each time. Pseudorandom number generators are also used to help ensure that sensitive information is transmitted securely by encoding it, protected by a special number called a key that is generated randomly and needed to decrypt the transmission at the other end.

But the giveaway is in the name: *pseudorandom*. These numbers aren't really random. After all, they can be predicted if you know the algorithm used to generate them. Moreover, the sequence of random numbers will, eventually, repeat. Our example above repeated after only nine numbers, whereas the ones used in most computer applications today won't repeat until more than 10 to the power 5,821 different numbers. That's a lot—it dwarfs the number of atoms in the visible Universe, for example, but it's not infinite. And it means that if you waited long enough, and kept comparing the sequence of generated numbers to what you've seen before, eventually you would see the generated numbers repeat. And then you could predict the outcome each time, without even knowing the algorithm.

We need this sort of randomness in our numbers when we use computers for science. We often want to simulate complicated systems like the weather, or the galaxies in the Universe, or the behavior of animals. But we never know all the details of some of what we're trying to simulate: we can only describe them with a probability distribution.

Randomness and Determinism

If you go for a walk along a rocky beach, you'll find sand, pebbles, and stones, from tiny grains to rocks the size of someone's head. No two rocks are exactly the same; they can differ in size, shape, material, color. The properties of any given rock that you might pick up seem *random*, unpredictable.

But if we knew where in the ocean, or along the shore, this rock started, and how it came to be carried by the currents, buffeted by other stones and worn away by water, we could, if we understood those processes well enough, predict its size and shape and its location here on the beach, today.

The understanding would have to be detailed and specific—not how erosion works in the abstract or in the ideal, but how it acted on this particular stone, throughout its geological history. This would not be a particularly interesting calculation, but it would be at least in principle possible, since the geological laws are *deterministic.*[2]

Determinism is the notion that there are some things that can be predicted exactly. Newton's laws of physics are deterministic: the total of the forces applied to an object, and its mass, will tell you its trajectory. In our beach example, we could program a sufficiently powerful computer with all those forces and masses, along with other laws of chemistry and water flow, to predict the properties of those stones on the beach.

Absent such a computer program, we can't make detailed predictions. Instead, the sizes and shapes of the stones appear random.

The seeming randomness of the stones' properties only comes about because we are ignorant of their history. Most of the time, this is what we really mean when we use the word "random" to describe some event: we don't know enough about it.

But there is usually some order to be discerned. When we talk about randomness, it is often more appropriate to consider collections of similar—but not identical—things. One rock's properties seem random because it is different from all the others on the beach. We can describe the sizes of the rocks by figuring out their distribution: how many small rocks; how many medium-sized rocks; how many big rocks. Such a distribution is in the same mathematical family as the probabilities that we have been discussing and obey the same rules.

To get our distribution, take all the rocks on the beach, or, more practically, a few bucketfuls from a small part of the beach, and sort them by size. You can then write down how many are between, say, 1 and 10 millimeters, between 10 and 100 millimeters, between 100 and 1,000 millimeters, and so on, until you have accounted for all the rocks you have chosen, from the smallest to the largest, in one or another size bin. If I made a chart of these measurements, it would look something like our die-roll bar charts.

In astrophysics, we use this concept all the time. We might know all the deterministic physical processes that describe the evolution of a cloud of gas over billions of years as it collapses down and evolves into planets surrounding a star that eventually becomes one of the white dwarves that so concerned Chandra and Eddington. But we certainly won't know all the details of the initial conditions of any individual cloud of gas. However, my colleagues and I might have a model precise enough to give us not the exact values, but a probability distribution for the properties of a cloud of gas at the start of its evolution. So if we want to be able to use computers to simulate such a system, we will need to use our computer's pseudorandom number generator to simulate the initial conditions for this problem. And perhaps we can simulate not just a single system of a star and its planets, but a whole set of them—often called a population, or an ensemble—each starting with somewhat different conditions, but compatible with—we usually say "drawn from"—the probability distribution of our model.

We can then use a computer version of the laws of physics to evolve each of these stars with their surrounding planets and compare them to what we have observed: not just individual objects, but the whole evolved population making up our galaxy today. This gives us an observational constraint not just on the laws of physics and of the evolution of stars, but on our model for the starting points, for that initial ensemble.

These grand metasimulations of entire populations of objects (in this case, stars and their surrounding planets) are often called "Monte Carlo simulations," named after the town in Monaco along the Mediterranean Côte d'Azur, again highlighting the link between probability, randomness, and the casino for which Monte Carlo is especially known.

And just as with the outcomes of our dice rolls, we can make bar charts of the observed brightness of the stars in the Milky Way and compare them with the output of our many simulations. And how do we finally compare our model and our data? This, finally, brings us back to Bayes' theorem, which allows us to choose among competing models that could explain our observations.

"Grocid Pondenome of Demonstures"

We saw that taking a probabilistic view of language allowed us to compress words and sentences using Morse code or zip files. But what if we take this model more seriously?

Claude Shannon was a twentieth-century American scientist, engineer, and mathematician and one of the first to look at language in this probabilistic way. He built up a series of models of the English language from very simple building blocks.[3]

First, he asked, what if I randomly choose letters (and a space to mark the end of a word) from a uniform probability distribution—the same probability (1 in 27 for the 26 possible letters plus one more for the space) for Z and Q as for E, T, or a space. When he did that, the letters he got were

XFOML RXKHRJFFJUJ ZLPWCFWKCYJ FFJEYVKCQSGHYD QPAAMKBZAACIBZLHJQD.

(This and the following examples are from Shannon's original paper.) These combinations of letters do not even rise to the level of senseless gibberish.

Next, he chose letters according to their frequency in some large set of texts (luckily for Shannon, such tables had been painstakingly compiled by hand and published in the early twentieth century). Shannon's random algorithm gave him

OCRO HLI RGWR NMIELWIS EU LL NBNESEBYA TH EEI ALHENHTTPA OOBTTVA NAH BRL.

The lengths of the words are about right for typical words in English, and there are more vowels used in his string than consonants, but still this sequence is not even pronounceable.

Now, he made his model more complicated: rather than just look at the frequencies of individual letters, he considered the frequencies of pairs of letters together: "th" was more common than "ur," and both of those pairs were much more common than "wa" or "xe"; and of course there are some combinations that never occur, like "qd." But this model doesn't just pick a series of pairs independently. Instead, he used something called a "Markov chain," named after the Russian mathematician Andrey Markov. The model looks at the last letter chosen, then picks the following by choosing among all the pairs starting with that letter, using their relative probabilities. So, if the last letter chosen is *t*, the most probable next letter is *h*, which is about

4 times as likely as *i*, which is itself only a little more likely than *e*, followed by the other 24 possible letters, as well as a word-ending space. Let's say *i* is chosen; the procedure looks for the probabilities for finding all the possible letters to follow it. Doing this, Shannon generated

ON IE ANTSOUTINYS ARE T INCTORE ST BE S DEAMY
ACHIN D ILONASIVE TUCOOWE AT TEASONARE FUSO
TIZIN ANDY TOBE SEACE CTISBE.

This sequence seems closer to words in the English language: most of the "words" are pronounceable, but of course most of them aren't really words, or even built up out of sensible parts.

But Shannon didn't stop at just considering pairs of letters. He also looked at triples, choosing the next letter based on the previous two in the sequence. Based on that, he produced

IN NO IST LAT WHEY CRATICT FROURE BIRS GROCID
PONDENOME OF DEMONSTURES OF THE REPTAGIN
IS REGOACTIONA OF CRE.

This sequence is even closer to English; there are some real words here, and if I saw "pondenome" or "demonstures" in something I was reading, I might look them up. But again, Shannon saw that he could build a better model. Rather than just consider longer and longer sequences of letters, he considered the frequencies of words. Again, he started by choosing them according to the frequencies in some known texts, getting

REPRESENTING AND SPEEDILY IS AN GOOD APT
OR COME CAN DIFFERENT NATURAL HERE HE THE
A IN CAME THE TO OF TO EXPERT GRAY COME TO
FURNISHES THE LINE MESSAGE HAD BE THESE.

Now, these words are real, and more common words occur more frequently, but since the model chose them independently, any sense of grammar is lost. But Shannon continued to play the same Markov chain game, always choosing the next word by its probability to follow the previous one. The model produced

THE HEAD AND IN FRONTAL ATTACK ON AN ENGLISH
WRITER THAT THE CHARACTER OF THIS POINT IS
THEREFORE ANOTHER METHOD FOR THE LETTERS
THAT THE TIME OF WHO EVER TOLD THE PROBLEM
FOR AN UNEXPECTED.

Now, each pair of words is sensible English, but of course any sense of coherence and grammar is lost very quickly.

With these increasingly complicated models for language, Shannon sought out a quantity that could describe the information content of some text: How could he encode the departure from randomness of a sequence of letters or words?

Shannon arrived at the idea of entropy, a quantity that depends only on the probabilities of the different symbols that make up the chain. This concept gives us a choice: we could consider the probabilities of different letters, considering each of them a different symbol, appropriate to Shannon's first four models. Or we could consider the entire words as symbols, as appropriate for his final two models.

Once this choice is made, Shannon's entropy links directly to one of our definitions of randomness: the entropy of a process describes precisely how compressible an arbitrary sequence from that process will be. For any set of symbols, the maximum entropy occurs when each one has the same probability—that is, when the principle of indifference holds. As we saw, these sorts of sequences are the most random, the least predictable, and hence the least compressible and the least likely to be understandable. Any sort of regularity or redundancy will decrease the entropy and increase the predictability and hence the compressibility.

But even with Shannon's most sophisticated models, considering triples of letters or pairs of words, the outcomes are not particularly close to strings of the English language. Perhaps this Markov chain idea is not a good model for language? Or perhaps Shannon didn't go far enough—why stop at only pairs of words?

Given the meagre technology available to Shannon when he did this work in the 1940s, he had to generate all of these models without the help of a computer, and pairs were all he could handle. Today, however, a program to generate these sorts of sentences fits in about 30 lines of code.[4]

And we can go still further: rather than look at the last one or two words to determine the probability of the next, we can consider the last several thousand words. This is the basis of the so-called large language models such as ChatGPT that seem to be able to at least mimic human language.[5] Rather than only considering the previous word, ChatGPT-3 considers the previous 4,096 words, and its successor, ChatGPT-4, increases this to 8,192 words.

Even 75 years after Shannon created these models, large language models like ChatGPT are computationally challenging problems. They begin with, well, the internet. Searching through online text, supplemented by specific sources such as Wikipedia and digitized versions of books, ChatGPT uses a "corpus" of something like 500 billion words, equivalent to about 6 million books the length of the one you are reading now. (For comparison, the US Library of Congress houses about 20 million books in English.)[6] But that isn't nearly enough to ensure that every possible intelligible sequence of 4,096 words will occur. This is where an AI model like ChatGPT differs significantly from Shannon's model. Rather than just recording the frequency of finding a given sequence of words in a set of texts and using those frequencies as probabilities to generate subsequent words based on the previous ones, these models need to work out probabilities even for sentences and phrases that have never appeared in any written text.

This so-called training step starts with a very general model, which allows the computer program to calculate probabilities even for those combinations of words that it has never encountered. The model, called a neural network, is intended to mimic the way that neurons are connected in a brain, and the training feeds the text to the network, letting the program build up connections based on that input, strengthening the connections when they make correct predictions and weakening them when they don't. The same sorts of models have become common in computing over the last several years: they identify faces in your smartphone's photo library, are used by scientists to speed up mathematical calculations by factors of hundreds or thousands, and are used by social networks to help predict your susceptibility to different kinds of advertising.

However, despite the suggestive naming of the neural network, which reminds us of our own brains and minds, don't be fooled: there is no reason to suspect that our neurons learn about language in the same way as these models. You didn't need to hear or read nearly as many combinations of words as ChatGPT to become proficient at your native language. Nonetheless, the connections built up by these models do, somehow, embody the meaning of the words and concepts.

When I asked ChatGPT-3.5, "What is Claude Shannon most famous for, and what does it have to do with you?," it responded:

> Claude Shannon is most famous for his pioneering work in the field of digital circuit design theory, as well as for his foundational contributions to the field of information theory. He is often referred to as the "father of information theory."
>
> . . . As for the connection to me, ChatGPT, Shannon's work on information theory is relevant in the sense that the principles he established have influenced the development of artificial intelligence, natural language processing, and machine learning. Concepts like entropy, coding theory, and efficient data representation have been applied to various aspects of AI and language models like me. Information theory provides a foundation for understanding how information is processed, transmitted, and structured, which is crucial in designing algorithms that power language understanding and generation tasks.

EIGHT

Entropy and Matter

The concept of entropy did not originate with Shannon and his efforts to understand the mathematics of communication. Rather, in the middle of the nineteenth century, physicists tried to come up with a modern theory of heat, superseding the old idea that it was a substance, "caloric," that flowed between objects and thereby controlled their temperature: more of this caloric meant that a body would be hotter.

The new idea was called the "mechanical theory of heat," which took its inspiration from the steam engines then accelerating the industrial revolution. This technology came from the long-known experimental fact that most substances, especially gases like water vapor or air, typically expand when heated. Heat was now seen to be related to the ability to do mechanical work—for example, to use this expansion to move a piston in a combustion engine.

Any physical substance, and in particular the gases that were of interest to the engineers and physicists studying the principles behind these engines, could be described by a small set of numbers—temperature (how hot it is), mass (how much it weighs), volume (how much space it takes up), and pressure (how hard it pushes on the walls of a container; this one is a bit more obscure, but obviously useful when you are trying to build a machine that pushes things around)—along with a description of the kind of substance it is (air, steam, or the individual gases such as oxygen or nitrogen that were just being isolated at that time). By the middle of the eighteenth century, scientists had discovered that these quantities were related to each other: if you compressed a gas so that it took up less space, but kept the temperature the same, it would increase the pressure. If you warmed up the gas in a rigid container, it would increase the pressure, or if you warmed it up in a container with a lid, it would expand—this is how you build an engine. These relationships weren't just vague tendencies though; they could be expressed as precise mathematical laws, known as equations of state.

Think about the air in the room you are in right now. You are probably within a few thousand feet of sea level, and the temperature is somewhere around 20 degrees Celsius (or 68 degrees Fahrenheit). For most purposes, these quantities are all that you need to describe the conditions in the room.

We also know that air is composed of a myriad of individual molecules, a mix of mostly nitrogen and oxygen. A typical room might have a billion billion billion molecules all together—that number is 1 followed by 27 zeroes. We couldn't possibly catalogue the locations, speeds, and directions of each one.

But the equation of state seems to tell us that we don't need to know all those numbers; all that matters are the numbers that go into the equation of state. This is an incredible amount of data compression: from a billion billion billion numbers to just five, all related by a single mathematical formula.

Engines and Entropy

The German physicist Rudolf Clausius, building especially on work by the French father and son Lazare and Sadi Carnot, showed that you could understand the equation of state by making a distinction between the temperature of an object—measured by a thermometer—and the amount of heat that it absorbs in changing its temperature, and moreover it was exactly this transfer of heat between two substances that allowed the steam engine to work. "Work" has a technical definition: it is the amount of force that the engine can supply, multiplied by the distance over which it can push ($W = F \times d$), so you increase the amount of work done by either pushing harder, or pushing for a longer distance.

Work is a kind of energy, and so is the heat transferred between substances. Clausius realized that it made sense to define a third kind of energy that was intrinsic to the specific object or substance, simply called the "internal energy," which is the energy stored within an object and depends on such properties as its temperature, mass, and composition. With this quantity, Clausius was able to define what we now call the first law of thermodynamics: the internal energy of some object is increased by the amount of heat it absorbs and decreased by the amount of work that it does.

This law is just a version of the broader concept of the *conservation of energy*: energy can be converted between the different forms of heat, work, and internal energy, but cannot be created or destroyed.

The most efficient engine would therefore use the maximum amount of heat with minimal change in internal energy, converting as much heat as possible to work. Unfortunately, Sadi Carnot and Clausius had realized that there are limits on the efficiency of an engine beyond the conservation of energy as demonstrated in the first law: you cannot make an engine that converts heat completely into work. After heating the gas that moves the piston, the gas needs to cool down to its original temperature, and the piston needs to return to its starting position. We can then add heat again, which moves the piston to do useful work, and repeat the cycle.

Carnot envisioned an idealized version of such an engine and discovered that even it could never convert all the heat energy into work. His engine consisted of a single piston being pushed by a gas and two separate "reservoirs," one hot and one cold. In a steam engine, reservoirs are containers of hot or cold substances large enough that their temperatures do not change even when they are used to cool or warm the piston's gas. Energy flowing into the gas from the hot reservoir pushes the piston up and heats it; energy flowing back to the cold reservoir lets it return to its starting position and cools it down. The maximum efficiency is determined by the ratio in temperature between the hot and cold reservoirs: a greater difference—hotter hots or colder colds—gives a more efficient engine. Carnot showed that any real engine would have to be less efficient than this.

Trying to understand Carnot's results, Clausius needed to define another new quantity beyond the internal energy of the gas. He called it "entropy," a name purposely reminiscent of the word "energy": it was related to the amount of energy *unavailable* for useful work.

Clausius saw that this quantity, entropy, could in an ideal case stay constant—or increase for real engines—over a cycle like the one described above. If we look beyond the specific case of Carnot's engine, we can see this implied that the entropy of a so-called isolated system that is not in contact with anything else must increase.

In this more general setting, applied to any system, not just an engine, this concept of entropy became known as the *second* law of thermodynamics: entropy only increases.

One of the simplest repercussions of the second law is that heat flows from warm to cool, not the other way around, without adding yet more energy to the system. Nearby objects tend to become the same temperature, a hotter object cooling down and a colder object warming up. This is known as reaching *equilibrium*, but it is also why ovens and refrigerators can heat and cool their contents and why your cup of coffee cools down eventually to the ambient temperature if you don't drink it.

What does Clausius's entropy, a bookkeeping device to help account for the inefficient conversion of heat into work, have to do with Shannon's entropy, defined in terms of probabilities?

The answer came only a little later in the nineteenth century, from Ludwig Boltzmann in Austria. He, along with J. W. Gibbs in the USA and James Clerk Maxwell in the UK, showed that entropy could be understood using a molecular or atomic model of substances. Today, such a model, in which substances are made up of vast numbers of microscopic particles, is obviously correct; at the time, these individual atoms and molecules had not yet been observed. But the model enabled Boltzmann and his colleagues to think about the relation between the positions, speeds, and directions of the individual particles making up a gas, and its large-scale properties such as internal energy and entropy as well as more directly observable quantities like the pressure, temperature, and volume.

Maxwell was one of the first to think about the different numbers of particles in different states. In particular, he worked out how the distribution of the velocities of particles in a gas depends on the temperature of the gas. He discovered a detailed formula, mathematically equivalent to the bell-shaped or Gaussian distribution that came up in our very different discussion of error bars earlier. This meant that the average speed of a particle increases with temperature: a hotter substance has particles with higher speeds.

How do we connect this small number of macroscopic quantities to the vast number of possible states of all the particles together? Remember that the air in your room has a billion billion billion particles, and each of them has a position, direction, and speed. Boltzmann, Maxwell, and Gibbs saw that the important quantity was not that very large number of particles itself, but rather the probabilities of each one, which was enough to tell us all the large-scale properties of the substance in which they were contained. Boltzmann and Gibbs, in fact, wrote down an entropy formula identical to the one Shannon would use three-quarters of a century later.

This formula depends on the individual probabilities of each possible state: how likely each specific particle is to be in a particular place, moving at a particular speed, in a particular direction. This could be overwhelming to calculate, in principle depending on each of a billion billion billion particles.

To make this formula work, Boltzmann had to make an intuitive leap: What if each possible state of the substance—every possible set of values for those billion billion billion particles—was equally likely? Then, we can apply our old principle of indifference and assign equal probabilities to each of those states. If there are W possible states, then the principle of indifference dictates that each probability will be 1/W.

With this vast simplification, Boltzmann in 1876 wrote down a formula for entropy that only depended on two numbers: Boltzmann's constant and the number of possible states in a system. (Boltzmann's constant is used for converting between entropy, temperature, and energy, but is an artifact of the units we have chosen to describe each of them.) The formula is:

$$S = k \log W.$$

In this formula, S is entropy, k is Boltzmann's constant, W is the number of possible states, and log is the logarithm function from mathematics. The properties of the logarithm also mean that Boltzmann's entropy is always greater than zero.

Boltzmann died by his own hand in 1906, before his formula could be experimentally verified and before it could be completely understood in terms of the coming theory of quantum mechanics in which his counting of states would make more sense. His entropy formula, in honor of its importance to the physical theories that have come since, is inscribed on his tomb in Vienna.

Boltzmann's equation is remarkable because it is so counterintuitive. If the entropy of an object only depends on its number of states, then it does not depend on the details of those states. It does not depend on any interactions between the billion billion billion particles that compose it (interactions that we know, in fact, to be present). How can this be so?

The answer comes back to that enormous number, the billion billion billion particles that we typically have to deal with. When there are so very many particles, and hence so very many possible states for an object to be in, the probability functions are subject to a remarkable mathematical phenomenon called "concentration of measure," a result originally proved by Shannon himself. This concentration of measure means that in systems with such large numbers of particles, the details of the probabilities become totally irrelevant, and the more detailed sum-over-probabilities formulae that Gibbs and Boltzmann—and, later, Shannon—wrote down just become equal to Boltzmann's elegant equation.

Concentration of measure is the reason your carbonated drink makes a satisfying hiss when you open the can, and why you can't get the carbonated gas back in, or why boiled-off water never recondenses back into an uncovered pan, or why my kids' toys inevitably end up spread around the house. Matter is a really good generator of randomness.

"We Can't Win"

In his science fiction novel *Do Androids Dream of Electric Sheep?*—much more famous as the basis for the movie *Blade Runner*—Philip K. Dick coined the term "kipple," described in this conversation between J. R. Isidore and his new neighbor, Pris Stratton:

> "Kipple is useless objects, like junk mail or match folders after you use the last match or gum wrappers or yesterday's homeo-

pape. When nobody's around, kipple reproduces itself. For instance, if you go to bed leaving any kipple around your apartment,
when you wake up the next morning there's twice as much of it. It
always gets more and more."

"I see." The girl regarded him uncertainly, not knowing
whether to believe him. Not sure if he meant it seriously.

"There's the First Law of Kipple," he said. " 'Kipple drives out
nonkipple.' Like Gresham's law about bad money. And in these
apartments there's been nobody here to fight the kipple."

"So it has taken over completely," the girl finished. She nod-
ded. "Now I understand."

"Your place, here," he said, "this apartment you've picked — it's
too kipple-ized to live in. We can roll the kipple-factor back; we
can do like I said, raid the other apts. But—" He broke off.

"But what?"

Isidore said, "We can't win."

"Why not?" . . .

"No one can win against kipple," he said, "except temporarily
and maybe in one spot, like in my apartment I've sort of created a
stasis between the pressure of kipple and nonkipple, for the time
being. But eventually I'll die or go away, and then the kipple will
again take over. It's a universal principle operating throughout the
universe; the entire universe is moving toward a final state of total,
absolute kippleization."[1]

Kipple is entropy, and "kippleization" is the heat death of the Universe,
everything eventually in equilibrium with everything else. (By mentioning
a "First Law of Kipple," Dick was lampooning and/or honoring both the actual laws of thermodynamics and, probably, another numbered set of science fictional laws, Isaac Asimov's three laws of Robotics.)

Of course, with apologies to Dick, what's true at the microscopic level of
molecules in motion isn't strictly the case for the crap that collects in your
house. You can, if you try, tidy up, but as we all know, it's hard to keep things
that way: there are a lot more ways for your house to be untidy than for it to

stay neat—just as with probability and Boltzmann's concentration of measure, it's much easier to make a mess than tidy up. There are many, many ways for a room to be messy but really only one way for it to be completely tidy. My kids can open a single cabinet, and within five minutes an entire floor of our house is strewn with toys, crayons, paper, and ribbons. Eventually, and much more laboriously, it will all be put back in something resembling its initial location. But it won't happen naturally: it requires cajoling, snack-bribes, threats, direct parental involvement. Indeed, left to their own devices, the kids will, of course, just make it worse.[2]

Every individual movement, of a toy from the shelf, or of a gas molecule moving from inside the can to outside, looks as if it could be easily reversed. But as there are more and more toys (it feels like dozens, and that's plenty), or more and more molecules (back to that billion billion billion number), it gets harder and harder to reverse all their trajectories so precisely that they would end up back where they started.

Even more generally, it's very rare for a system with so many constituents to ever be in a situation that is very far from average. Boltzmann's equation says: if there are more possible states, so W is larger, then the entropy of the object is larger. If there are more possible states for an object to be in, we have less information about the system, since we don't know which one of this larger number is the correct state. A system with higher (thermodynamic/Boltzmann) entropy is more random, just as a probability distribution with higher (information/Shannon) entropy is more random.

And there are many ways to change the number of possible states for an object to be in. We can change the volume of a gas: a larger volume means more possible ways for the same number of particles to be arranged. We can raise the temperature: Maxwell showed that a higher temperature is equivalent to faster particles, and so more possible speeds are available to each particle. Or, we can combine two or more systems into one: the total number of possible states for the two systems comes from multiplying the number of states of each individual system, vastly increasing the total number. Then, due to the mathematical properties of the logarithm function, this in turn means that the total entropy of the combination of two systems comes from adding up the entropies of the individual systems. If the subsystems are in contact, heat can flow between them, bringing them to equilibrium, which raises the entropy still further. My cup of coffee might be 190 degrees Fahrenheit, and my pitcher of milk only 40 degrees. When they mix, they will eventually get to a temperature somewhere in between (depending on not only the two temperatures but also the quantity of each). But the single mixed white coffee has many more possible states altogether than the cup full and dollop from the pitcher it began with—it includes all of the original separate possibilities, along with the much more likely ones in which they are actually mixed.

But to really understand entropy, from Clausius's point of view, and Boltzmann's, and Shannon's, we need to return to our slogan: all probabilities are conditional. Not all models are relevant to our inferences. Boltzmann and Shannon have shown us that we will assign essentially the same probability to each state of an object even if we know that there are forces in between the particles of that object, so we may as well ignore them. Given the concentration of measure theorem, we have to put in a lot more information to make a difference.

Maxwell himself put this in an evocative way. Think about a refrigerator, which is kind of the opposite of an engine: it uses mechanical work (some sort of pumping system) to cool its contents rather than heat them like an engine does. In this case the second law of thermodynamics says that the surroundings of a refrigerator will heat up and that the average temperature of the refrigerator and its environment will increase—heat cannot flow from a hotter substance to a colder one without doing work.

So Maxwell imagined a box of gas with a "demon" (a word first used by Kelvin describing Maxwell's idea) in control of a microscopic door in a wall running down the center of the box. If the demon could observe the speed of each particle, it could open the door and let the fastest particles through in one direction and the slowest particles through in the other direction. Eventually, this would lower the average speed—and hence the temperature—of the particles in one half of the box and raise the speed and temperature of the particles in the other half. This would be a perfect refrigerator, cooling half of the box with no input of work. But this system has lower entropy than the original one at a single temperature, seemingly in violation of the second law of thermodynamics.

That is, after the demon has done its work, the gas in the box has a lower entropy—if we stick to Clausius's definition of entropy, which depends only on its large-scale properties. But in Maxwell's imagined experiment, the demon has a lot more information than just the temperature and volume of the gas: it knows the detailed position and speed of all or most of the particles inside, at least when they get near the door. If we could write down the probabilities for those particles' positions, speeds, and directions, they would be very different, conditional upon this new information. And this is exactly the situation in which the concentration of measure theorem doesn't work—Boltzmann's simple and celebrated expression no longer applies.

There is another way to look at Maxwell's experiment—we can step outside the box and consider the demon itself. After all, it too is part of the complete system. What contribution does it make to the entropy? Landauer's principle, developed by Rolf Landauer while working at IBM in 1961, says that manipulating information takes energy, just like moving a piston. And just as Carnot's engine is limited by the need to return to its starting point—to complete the cycle—any information stored eventually needs to be erased. Landauer showed that this erasure must add heat—energy that is not immediately available to do work—to the environment, increasing the total entropy.

This is one of the most direct links between the thermodynamic entropy of Clausius, Boltzmann, and Gibbs to the information entropy of Shannon. When the demon erases its information about the particles of gas—when it forgets—the information entropy of its probability distribution for those particles will increase, and so will the overall entropy of the system due to Landauer's principle and the heating of the environment.

Almost any useful process decreases the entropy somewhere: it makes order out of disorder. The second law tells us that the process—using a computer, running an engine, cooling something with a refrigerator (or even with a demon!)—will give off heat that warms the environment. This is no less true of natural processes that may seem to create organized systems out of their constituent parts.

Stars are formed from lumps of gas slightly denser and heavier than their surroundings. This increases their gravitational attraction, making them contract and pulling in yet more material. As a star contracts it warms up, and as it warms it starts to glow, and eventually it ignites its nuclear furnace. A star itself is smaller and more compact—more ordered—than the gas from which it formed. The gas considered alone has lost entropy. But the star is shining, and its starlight serves to warm up the environment— increasing the total entropy.

The evolution of life itself seems even more puzzling than the formation of physical systems like stars: a planet teeming with cells, and then plants and animals, is far more ordered than the undifferentiated primordial ooze present on Earth before the first organisms. But the chemistry of life converts food into movement, both on a microscopic level as the different parts of cells are assembled from their building blocks, and on larger scales within organisms (muscles, the flow of blood) and organisms themselves moving within their habitats. All this movement is due to tiny molecular engines, each subject to the second law, dumping waste heat into their environments. The amazing local organization of living creatures is at the expense of increased global entropy and a warmer environment.

NINE

Quantum Randomness

The so-called classical physics of Isaac Newton's mechanics, James Clerk Maxwell's electromagnetism, and Albert Einstein's gravity makes specific predictions: if you give me enough information about the locations of the heavenly bodies, I can predict their future paths (although it may take me vast amounts of computing power to do so). Maxwell's demon may be unrealistic, because the idea of dealing with information about a billion billion billion particles seems fanciful—but not logically impossible. Probabilities appear in the science of thermodynamics only because we don't have complete information.

Probabilities are built into *quantum mechanics* from the start.

Quantum mechanics determines the probabilities that we should assign to the possible outcomes of our measurements: How likely is an electron to go through the opening of our detector? How likely is the nucleus of an atom to decay in the next ten minutes? How likely is an electron to be spinning clockwise vs. counterclockwise?

The mathematical rules of quantum mechanics are simple. Most of what we care about is encoded in something called the *wave function* (often denoted by the Greek letter psi, ψ), similar to the probability functions that we have worked with before, like $P(x)$, the probability that a star is at some distance x. The wave function depends on observable properties of a particle such as its position or its velocity. Sometimes, the wave function is called the "quantum state," supposedly encoding everything that there is to know about whatever we are studying. When an observation has more than one possible outcome, we say that the wave function or state is in a *superposition* of the different possibilities of states. "Superposition" is basically a fancy word for adding up: the wave function is the sum of the individual wave functions that encode each of the possible outcomes.

If we know about the forces acting on the particle, the rules of quantum mechanics then give us an equation—the famous Schrödinger equation—we can solve to work out the wave function and how it changes with space and time. The wave function produces not a single number, but two, and we can combine those two numbers to work out quantities like the probability that a particle is in a particular position and how this probability might change with time. (For readers who are mathematically inclined: the wave function is a *complex* function, and we work out the probability by taking its squared absolute value, which is equivalent to squaring and adding up the two real numbers that compose the complex number. So mathematically we call this pair of numbers the single, but complex, value of the wave function, from which we can calculate our probabilities.) The formula for going from a wave function to a probability is called the Born rule, after Max Born, one of the early pioneers of quantum mechanics.

That probability might tell us how likely a particle is to be at a particular position, or have a particular speed, or be rotating around a given axis at a particular rate. A celebrated example from the early days of quantum mechanics was to work out the properties of an electron orbiting the nucleus of an atom. If the theories of Newton and Maxwell still applied in this situation, atoms would be like miniature solar systems—with the electric force taking the place of gravity—and we could use similar mathematical techniques to understand their behavior.[1] But quantum mechanics makes very different predictions—predictions in accord with our observations.

The most striking of these predictions, and one that gives quantum mechanics its name, is that the possible orbits of an electron around the nucleus are discrete. By looking at the light that is emitted by an atom, we find that the light comes out with only specific colors. (These colors appear as the same kinds of lines that are observed in the spectrum of light from the sun and that allow us to determine something about which elements are present there.) But the color of light tells us its wavelength or, equivalently, its energy. So, the energy of an electron in an atom is *quantized*.

This quantization arises because Schrödinger's equation for an electron in an atom only makes sense—it only has solutions—for specific values of the electron's energy. Each possible energy gives a different orbit, with higher energies roughly corresponding to being closer to the nucleus of the atom. Quantum mechanics makes a definite—confirmed—prediction: the energy of electrons in atoms will be quantized, at specific levels that can be calculated for different substances.

Despite these definite predictions, the wave function still represents probabilities. Schrödinger's equation says that the energy levels will be quantized, but in fact we can still solve the equation with *combinations*—superpositions—of these discrete energy levels, in combinations that will change over time. The specific way that the solution combines those energy levels depends on the state of the atom: how it has interacted with its surroundings, and any actions that we might have taken to prepare it for an experiment we might want to perform.

Then, the wave function gives us a recipe for calculating the probability that we will find a particular electron in a particular energy level at a particular time: work out the (complex) value of the wave function for that energy, and the probability is its absolute value squared.

But not everything in quantum mechanics is quantized. I can also use the wave function to work out the probabilities for the location of an electron, or its speed. The recipe is the same: evaluate the wave function for any location that I care about, and its absolute value squared gives the probability of finding the electron there.[2]

I can only check whether these probabilities make sense by doing an actual measurement. I am more likely to find energies with larger probabilities; similar for positions or velocity. But once I've made such a measurement, things have changed: the probabilities that I've calculated don't apply anymore. Now, I *know* the value of the energy or position or velocity. So the probability should be *one* for whatever value I have just measured—but that probability is represented by a different wave function, equal to zero at all the energies, or positions, or velocities, where I didn't find the particle.

This abrupt change of the quantum state when a measurement is made, shown in Figure 13, is called "collapse." Before measurement, the wave function

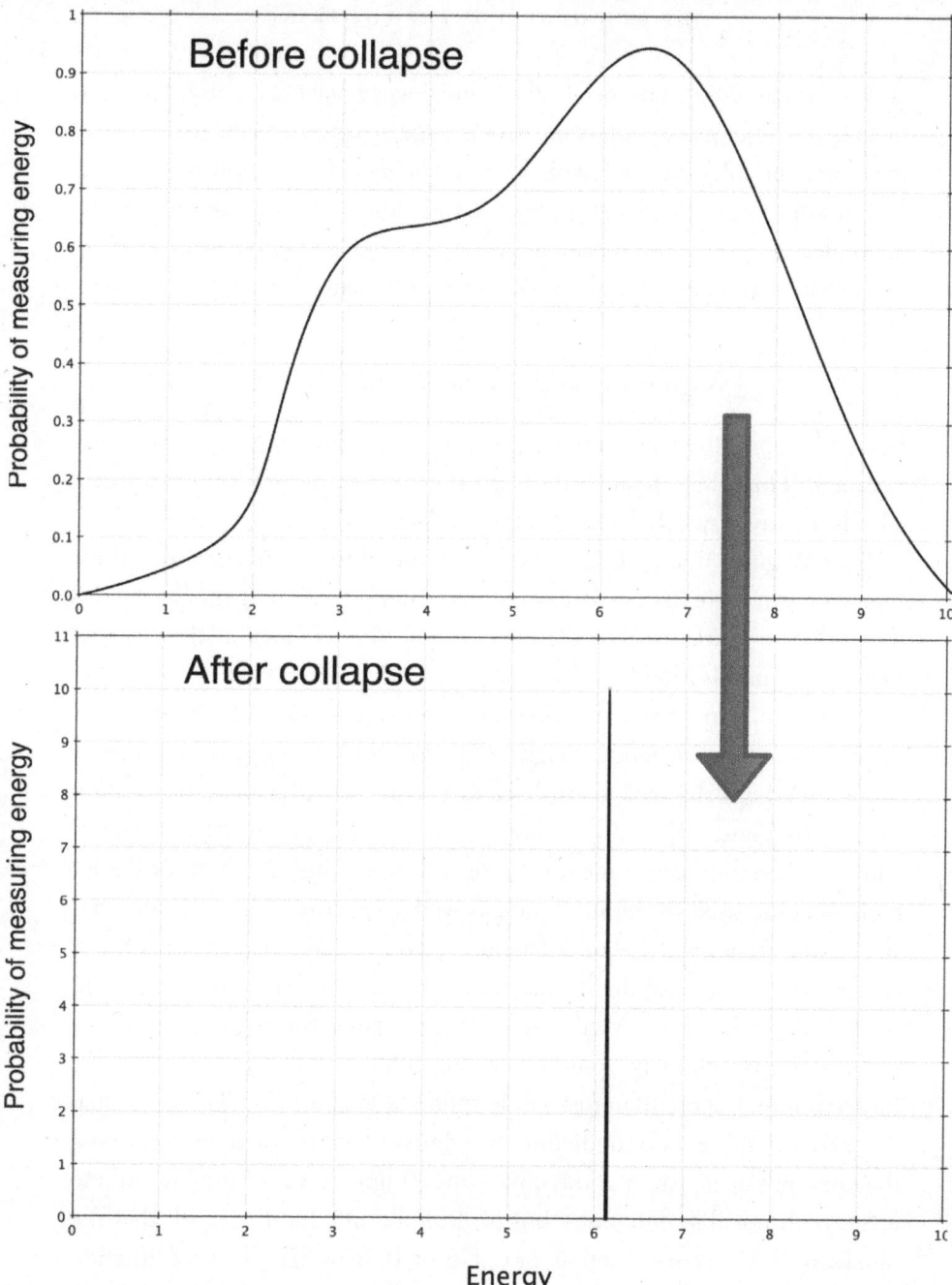

Figure 13. Energy wave function (squared) before and after measurement. In this example, the wave function has collapsed to a point near (but not exactly at) the peak of the initial probability.

evolves with time as determined by the Schrödinger equation, but just at the moment of measurement, it instantaneously changes to agree with the results of the measurement. After measurement, the Schrödinger equation again applies, but starting from a new state, with the energy (or position or velocity) fixed to its measured value. The quantum state of the Universe (or at least of the object that we are studying) abruptly changes due to our measurement.

Wigner's Friend and Schrödinger's Cat

Before collapse, the wave function records all possible outcomes of an experiment. Does that mean all those possible outcomes are, somehow, real? This is the mystery behind the parable of Wigner's friend.

Both Wigner (that is, Eugene Wigner, one of the pioneers of quantum mechanics in the early twentieth century, whom we already met as the author of "The Unreasonable Effectiveness of Mathematics in the Natural Sciences") and his friend are scientists. Wigner's friend is shut in her lab doing a quantum mechanical experiment, perhaps waiting for the decay of some radioactive substance. Wigner, in turn, is observing his friend from outside: perhaps she will report back to Wigner whether there has been a decay. To Wigner, this inner laboratory is just a larger-scale experiment, with a very complicated measuring device consisting of whatever equipment is being used in the inner lab, as well as the friend herself. But consider some time after Wigner's friend has completed her experiment, but before she has reported the results: for the friend, the wave function has already collapsed, but for Wigner, who does not know the result, it has not— it still describes the superposition of the different outcomes. If the wave function is real, more than just a description of Wigner's knowledge, or that of his friend, this is a contradiction: two different wave functions describing the same physical system at the same time. Wigner's wave function still encodes both possible outcomes, but his friend's only has the result that she obtained. Is this a contradiction, or is it just telling us that the wave function is about information rather than reality—just like conditional probability?

"Wigner's friend" is a thought experiment similar to, but less grisly than, the one involving Schrödinger's more famous cat. In Schrödinger's thought experiment, he places his cat in a box with a device that does some sort of quantum mechanical measurement. It might wait for the radioactive decay of an atom, or measure the rotation of some electron. Let's say that there are two possible results, each with a 50 percent probability of occurring.

Depending on the outcome, a machine in the box either feeds the cat her favorite food or administers a mild sedative. (As Schrödinger originally had it, in the second option the machine administers a deadly poison gas, but I can't quite bring myself to kill the cat, even in a thought experiment.) Until Schrödinger opens the box to look at his cat, the wave function encodes both possibilities and does not collapse until he learns the result.

These thought experiments force us to confront the meaning of the wave function in quantum mechanics. In the Wigner's friend scenario, two different wave functions seem to apply at the same time, depending upon whether we are considering Wigner's point of view or that of his friend. For Schrödinger's cat, the wave function exists in a superposition, with a cat who seems to be both awake and asleep[3] at the same time. (And indeed if we grant the cat enough sentience to understand its own situation, we would need to consider the feline point of view as well.)

In both thought experiments, we need to understand what it means for the wave function to describe more than one possible outcome before an observation has been made. We might be able to accept an electron in such a state, but it is much harder to understand what that wave function could mean for a more complicated object, and especially for a living being: the cat, the friend, or even ourselves. Is the cat somehow both asleep and awake? Has Wigner's friend somehow observed all possible outcomes of her experiment until she reports back to Wigner?

But even after Wigner and Schrödinger have observed their friend and cat, the wave function has not finished its job.

The Paradoxes of Quantum Uncertainty

The collapse of the wave function means that the probabilities for getting results from *subsequent* measurements depend upon the outcome of any previous measurement, or even whether we have already performed a measurement at all. On its own, this seems not just unobjectionable, but expected, returning to our slogan, all probabilities are conditional. Having the outcome of a first measurement and not having made the measurement at all are very different states of knowledge on which to base our probabilities.

But wave functions are not quite, or not only, probabilities, and their role in quantum mechanics gives them further predictive power through the details of how they change over time under the influence of Schrödinger's equation and then lead to probabilities through the Born rule. The structure of wave functions in quantum mechanics, and their relation to measurements and probability, leads to one of the most celebrated—and confusing—results of quantum mechanics, Heisenberg's uncertainty principle.

The uncertainty principle in quantum mechanics tells us that there are limits on how well you can simultaneously measure more than one property of an object or a system at the same time. The most famous example is the location of a particle and its velocity: the more accurately I measure one, the less well I can know the other one. The details of that limit are defined by the object's mass and Planck's constant, a new fundamental number that comes into any situation involving quantum mechanics.

If I could somehow know a particle's position perfectly well, I would have absolutely no information about how fast it was moving. If I could locate a hydrogen atom to a position of one ångstrom (one ten-billionth of a meter), the uncertainty principle says that I could only measure its velocity to an accuracy of about 600 meters per second—which is not far from the typical speed of a molecule of gas at room temperature. These are realistic numbers at these tiny scales.

But if we are not doing experiments in chemistry or particle physics, these limits rarely come into play: even if I could measure the position of a 1 gram object to an accuracy of 1 micrometer, the uncertainty principle states that I can only measure its speed to an accuracy of about 10^{-20} centimeters per second, far better than any realistic measurements I could possibly make. And it's similarly irrelevant for typical objects that we encounter, even those as small as a cell or a particle of dust, with yet smaller uncertainties in the speed as objects get more massive.

Even if the repercussions of quantum uncertainty are remote from every-day situations, the very idea of such a fundamental limit to our knowledge starts to seem different from the way we have used probability so far. The priors, likelihoods, and posteriors, linked by Bayes' theorem, were governed by de Finetti's even stronger slogan "probability does not exist," and hence reflects only our knowledge of the world. But in quantum mechanics, the laws of nature themselves seem to be telling us something about what probabilities we can apply, through Born's rule and the wave function. Quantum mechanical probability does seem to exist, independent of our knowledge.

Moreover, we need to use the quantum mechanical wave function to describe not only a single particle on its own, but also particles that can interact with each other. Any such interaction between particles can cause the wave functions for the particles to become "entangled." This is an evocative word, but what it describes is not much different than a concept that we encountered earlier: conditional probability. If I know the weather in Boston, I have at least some information about the conditions in New York. If you ask my daughter Olive about what she had for dinner last night, you can probably predict what her sister, Tallulah, ate.

Similarly, once two particles have become entangled, their fates, or our knowledge of their fates, become intertwined. If I make a measurement on one particle, I might have more information about the properties of the other.

Quantum entanglement is most famously seen when we consider a very specific property of elementary particles known as *spin*. In quantum mechanics, all fundamental particles have a fixed and immutable value of this property. For the particles that make up a physical object, this value is 1/2 — we say that electrons, neutrons, and protons are *spin-half* particles. Photons, the constituents of light, are *spin-one*. There are many other examples of both spin-half and spin-one particles. There is at least one *spin-zero* particle, the Higgs Boson, predicted in the 1960s and seen in particle accelerators in 2012. And we believe that there might be a kind of *spin-two* particle, the graviton, but its existence depends on the quantum mechanical properties of the gravitational force, still one of the great mysteries of physics.

We call this property "spin" because it is mathematically analogous to rotation, though we should not think of particles as tiny, rotating spheres. But as when we measure actual rotation, for spin we need to specify not only how fast something is spinning, but also in what direction. For spin-half particles like electrons we first decide what axis along which to consider the spin. And this is where the analogy with our classical conception of rotation starts to break down. For a top, say, we can look at it and ask about how it spins around its axis. Physicists have defined something called *angular momentum*, and we can measure this momentum along any direction we like, even away from the actual axis of rotation of the top—it will be largest along the axis, and usually zero perpendicular to it.

But spin-half particles are different. For *any* axis that we pick, we will get angular momentum of either +1/2 or −1/2 (measured in units of Planck's constant), and no other possibilities, very roughly analogous to the top spinning clockwise or counterclockwise along that axis. The probabilities of each of the two possible values are determined by the wave function. And as with position and momentum, there are limits on how well we can know the spin in one direction if we have measured it in another. Consider a simple case: I first measure the spin of the top along an up-down axis and then work out the probabilities for a measurement along a side-to-side axis, one that is perpendicular to my first measurement. In this case, no matter what results I get from the up-down measurement (+1/2 or −1/2, often called "up" or "down" in this case), Born's rule tells me that I will have *equal* probabilities for +1/2 and −1/2 along the new axis. If the second measurement is not perpendicular to the first, but at some other angle, the formula is more complicated, but the probabilities are still completely predicted by the results of the first measurement. In all cases, quantum mechanics tells us the probabilities, but not the actual outcomes.

We can set up a pair of particles—such as our spin-half electrons—so that they are entangled in a very specific way. We can arrange a system of two particles so that the joint wave function gives either particle a fifty-fifty chance (probability 0.5) of being either "up" or "down." But they are entangled in such a way that if one particle is measured to be "up," then the other will always be "down," and vice versa.

This was the thought experiment described by Albert Einstein, Boris Podolsky, and Nathan Rosen (a group usually called "EPR") in 1935, although this version (involving spin) is closer to a proposal by Aharonov and Bohm somewhat later.[4] They asked, what if we give one particle in the pair to each of two experimenters (often called "Alice" and "Bob") and move them far apart after setting up the entanglement? When Alice makes her measurement of the spin of her particle, the wave function collapses everywhere all at once: if she measures "up," we *know* that Bob will measure "down."

Alone, the outcome of this thought experiment does not seem like much of a problem. Imagine that I have a peach and an apricot, each in an identical box, and I give one each to Alice and to Bob, along with their entangled particles. When Alice opens her box to find a peach, we *know* that Bob will find an apricot. No mystery here.

But quantum entanglement is not a box of pitted fruit. Einstein's own theories of special and general relativity limit the way that an action at one place—Alice's measurement of her particle—can have an effect somewhere else—Bob's far-away particle. Relativity says that the concept of "all at once" (or "simultaneity") does not even make sense. Different observers moving at different speeds might see Bob make his measurement before Alice, rather than vice versa—we cannot even define the order of the two events.

According to EPR, this implies that the actual spin value that Bob would find must have already been determined, even if it is unknown. EPR argued that Alice's seeming ability to affect the outcome of Bob's experiment from afar means that the outcome—the value "up" or "down" that Bob eventually gets—must already exist in the world. In their terms, it must be an "element of reality." For that to be the case, quantum mechanics must be incomplete. There must be a "hidden variable," beyond the wave function and the Born rule, not part of the normal equations of quantum mechanics, that determines both Alice's and Bob's measurements. This, Einstein thought, would be the only way to reconcile quantum mechanics with his theory of relativity, avoiding the instantaneous collapse of the wave function—since "instantaneous" is a meaningless concept once relativity is considered.

In 1935, quantum mechanics was still young enough that such hidden variables seemed possible, although John von Neumann had published a proof in 1932 stating that no such theories were compatible with the mathematics of quantum mechanics as it was then understood.[5] In fact, this proof was flawed, but this flaw was not appreciated at the time.[6]

But a more serious counterargument was constructed by John Bell in 1966, in a paper explicitly called "On the Einstein Podolsky Rosen Paradox." Bell's argument relies on the fact that Alice and Bob could have, instead, measured their spins along different axes from one another.[7] For any pair of axes—directions for their respective measurement devices—we saw that quantum mechanics gives a formula for the probabilities of finding up or down (+1/2 or −1/2) in whatever direction each of them measures.

But we know that we can assign probabilities for any model that we choose. So Bell also asked what Bob's probabilities would be for a model that satisfies the following criteria:

- The values measured by Bob exist independently of their being observed—usually called "realism."
- Alice's measurement at a different location cannot instantaneously affect Bob's results—usually called "locality."
- Bob's choice for his measurement direction does not depend on Alice's choice for hers—this is called "measurement independence."

The realism criterion says that the world exists entirely separate from our observations of it: we don't have to perform a measurement for there to be a fact of the matter for the result—we just don't know what it is. The locality criterion says that causality only happens, well, locally: Einstein's theories, with their ultimate limit of the speed of light, should not allow what Einstein himself referred to as "spooky action at a distance."[8] The criterion of measurement independence says that not only do Bob's results not depend on Alice's, but neither does his choice of direction, or anything else about his setup that could have an effect on his measurement. This is sometimes called the "free will" or "no conspiracy" requirement: Bob has the power to *choose* what measurement he will make.[9]

These criteria seem unobjectionable, even desirable, in a scientific theory. For the example of the peach and apricot, they appear obviously true.

But these mild requirements do not agree with the predictions of quantum mechanics.

For the entangled electrons, Bell's theorem tells us that at least one of those requirements—realism, locality, or measurement independence— must not apply. That is, if we want there to be a reality that determines both Alice's and Bob's results, there must be something in the underlying physics transmitted faster than light.

That convoluted last phrase—"something in the underlying physics transmitted faster than light"—indicates the murky situation. Einstein's theories tell us that particles cannot move faster than light, but the fundamental physical principle is more subtle. We cannot send information faster than light. Even with Bell's results, this is still true.

Alice's result on its own is unpredictable—she will see up or down with fifty-fifty probability. Once she makes her measurement, she knows for certain what Bob will see. But as far as Bob is concerned, his result is unpredictable, too—unless she sends him information about his result (by email, carrier pigeon, text message, or just walking over to his lab, all of which obey Einstein's speed limit), he doesn't have any way to access Alice's result except for performing his own measurement. Only after his experiment will he know for certain what Alice has seen. So: we cannot use entanglement to send a signal from Alice to Bob or from Bob to Alice. Despite Bell's theorem and its claim that quantum mechanics might require nonlocal solutions, it is not clear whether, or to what extent, quantum mechanics is actually in contradiction with Einstein's theories: neither Bell's nor Einstein's theories allow sending information faster than light.

Shut Up and Calculate

These puzzles reflect a deeper confusion about what is usually called the "interpretation" of quantum mechanics. The rules for doing physics with quantum mechanics are straightforward: write down the wave function and use Schrödinger's equation to work out how it changes with time and use

Born's rule to make predictions for the probabilities of any measurements that you might make. A century of experiment and observation since Schrödinger, Born, and Heisenberg were working has never once shown a contradiction with these rules.

But mathematical formulae do not tell us what is *real*. Does the wave function exist, or is it just a mathematical tool, like the probabilities in Bayes' theorem, to help us learn about the world? What happens when we make a measurement? Do particles have both a definite speed and a definite position at the same time?

All the different ways to answer these questions have led to a zoo of different interpretations of the mathematical formalism of quantum mechanics, different ways of understanding how the equations map on to what exists—the ontology of the theory.

The interpretation implicit in my description of the laws of quantum mechanics, involving a sharp divide between the classical apparatus making the measurement and the quantum system under observation, resulting in the instantaneous collapse of the wave function when an observation is made, is usually called the "Copenhagen interpretation." The name arises from the group working around the Danish physicist Niels Bohr in the late 1920s, including both Heisenberg and Born. Indeed, while they and other visitors to the Institute for Theoretical Physics in Copenhagen (today known as the Niels Bohr Institute) formulated and formalized the mathematical structure of quantum mechanics, Bohr's contributions by this time were somewhat more philosophical.

Alongside the primacy of the wave function and the probabilities that it encodes through the Born rule, the Copenhagen interpretation also requires a split between the quantum and classical worlds: the quantum is the subject of the measurement, but the result itself—the digits on the screen of a device, or the marks on the paper where they are recorded—exists with classical certainty. Moreover, those probabilities mean that the world is fundamentally random and unpredictable. There is no information beyond the wave function itself—there are no hidden variables.

All of this does not quite point to the true underlying nature for the world—its ontology. Does the wave function "exist"? I am forced to use a pair of much-loathed scare quotes around that word "exist" because it is hard to know what it would even mean here. The wave function certainly describes something about the world, but it certainly does not exist in the same sense that an electron exists, or the chair you are sitting on, or you or me. It can't be poked or prodded, and you can't have a conversation with it.

Sometimes this confusion is taken as an admonition to "shut up and calculate"[10]—the mathematical rules of quantum mechanics are enough to give precise (albeit probabilistic) predictions for any possible circumstance, and we physicists should not worry about what lies beneath them.

This approach is how quantum mechanics appears in most textbooks: a set of mathematical rules and how to apply them to the world. It's how I learned the subject, and how I have taught it to second-year undergraduate students. The first chapter of the textbook I used when learning quantum mechanics presented a series of axioms, and the next 450 pages teased out their repercussions in the real world.[11] The difference between quantum mechanics and the axioms and proofs of Euclidean geometry is that the first chapter of Euclid's *Elements* gives axioms for defining such commonplace notions as points and lines, whereas the axioms of quantum mechanics require defining wave functions and even more obscure mathematical objects such as Hilbert spaces.

But one of those postulates seemed quite unlike the others: it states that to construct the mathematical objects describing anything we can observe, we start from the equations of classical mechanics—from Newton and his successors—and we follow a recipe to translate those equations into their quantum form. This is part of the "correspondence principle" developed by Niels Bohr as quantum mechanics was taking shape, eventually becoming a core tenet of the Copenhagen interpretation.[12] As part of the mathematical recipe, this makes sense, but it is harder to square with an understanding of the meaning of quantum mechanics. It seems to require that classical mechanics, the laws of physics that we apply to large-scale objects, is somehow more fundamental than the quantum mechanics that we apply to atoms and molecules—the particles that make up those very objects. This seems backward and leads students and scientists to choose to "shut up and calculate" rather than grapple with the implications of quantum mechanics' rules.

So, alongside the development of the physical repercussions of quantum mechanics over the last century—fundamental physics describing particles from quarks to the Higgs boson, and technology from computer chips to batteries and solar cells—scientists (and philosophers) have also been grappling with the so-called interpretation of quantum mechanics. We know what quantities the equations predict, but how can we understand what they mean about how the world is constituted?

Quantum mechanics has been taken up by philosophers as well as physicists over the last century. Heisenberg even said that "Bohr was primarily a philosopher, not a physicist."[13] But perhaps he was not a very good philosopher: Imre Lakatos called Bohr's own attempts to formalize quantum mechanics after about 1924 an example of a "degenerating theory"[14]—cooked up to agree with observations, but making no proper predictions.

Lakatos's friendly rival, Paul Feyerabend, the Austrian philosopher of science who championed an anarchic "anything goes" conception of the scientific method, discussed his point of view and his feud with Karl Popper:

> It remains to say a few words about Popper's contention that "the reduction of the wave packet . . . has nothing to do with the quantum theory: it is a trivial feature of probability theory." The "feature" which Popper has in mind is of course the dependence of probabilities on experimental conditions which entails that they change abruptly when the conditions change. Now the fact that the Born Interpretation is part of the quantum theory entails that this feature is part of the quantum theory also. . . . But while Popper, having before himself the ready-made result of the labours of the Copenhagen school, can regard the situation as a "trivial feature of probability theory," Bohr has shown how difficult it is to have both the quantum theory and a propensity interpretation of its predictions, and how many modifications are needed. . . . It is therefore much more complex than the simple "reduction" of propensities for which it makes room. Of course, there is no observable difference between the two reductions and an instrumentalist will feel no compunction in identifying them. But such an identification does little to advance our understanding of the quantum theory.[15]

(In Feyerabend's terminology, "reduction of the wave packet" is our "collapse," and by "propensity" he means the idea that there is some sort of physical mechanism through which the Born rule somehow enforces the way that the outcomes of experiments accord with their quantum mechanical probabilities.) This summarizes the questions and points of view that lead to the different interpretations of quantum mechanics' mathematical rules: Where do probabilities come from? What happens at the moment of measurement? Should we just use the mathematical rules of quantum mechanics and not worry about their interpretation?

Many Worlds

The most famous way of looking at quantum mechanics, one that has found its way into popular culture from comic books to the movies *Sliding Doors* and *Everything Everywhere All at Once* is the "many worlds" interpretation, first discussed by Hugh Everett in the late 1950s and updated by Bryce DeWitt in 1970.

As the Universe evolves, if the "many worlds" interpretation holds, everything that exists is described by a single wave function, changing over time according to the Schrödinger equation. That's it: there's no special moment of measurement, no collapse. In standard (Copenhagen) quantum mechanics, collapse prunes the wave function so that it only describes the—one and only—observed outcome of any measurement. The many-worlds wave function, without this pruning, still includes terms corresponding to every possible measurement outcome, of every possible measurement that might have occurred.

In the full wave function, each of those unpruned terms, often called a "branch," remains. How do we connect this enormous wave function to the physics of our experience? The proponents of the many-worlds interpretation say that each of those terms is a *world*, or really, a universe, constantly branching off as particles interact and become entangled with one another. In the EPR experiment, the branches split from one another at the start of the experiment, when the up/down (or down/up) pair are first created. In one branch Bob sees his particle's spin as up and Alice sees hers as down; in the other, those outcomes are reversed. Both branches exist, and they are created not by measurement but by the quantum mechanical process that randomly assigned the pair of particles their spins. The split happens at that moment of interaction, but until the measurements are made, the copies of Alice and Bob don't know which branch they are in and are otherwise identical until then: each copy of both Alice and Bob would assign the same fifty-fifty probabilities to their as-yet-unknown results. For each of them, the only collapse is in their Bayesian probability assignment, from fifty-fifty probability to 100 percent certainty about the result, no different than looking at a flipped coin to discover whether it has turned out heads or tails.

Within a world (or a universe, or branch) all the locations and possible measurement outcomes are fixed. Probabilities are still calculated by the Born rule, arising because we don't know which branch we are in until we look at the results of an experiment. Now, our measurements really do reveal a preexisting fact—describing the branch in which we happen to find ourselves, embedded in the grander full wave function describing the multiverse of all the possible branches existing alongside each other.

In the many-worlds multiverse, there is no need for that tricky moment of measurement and collapse that seems to require a crossover between a classical world of our measuring devices and the quantum world being measured. Without collapse, the moment of measurement loses its special place; branching can occur whenever *any* particles interact with each other, a roiling sea of comingling particles, every quantum contact potentially producing a new branch, as in Figure 14. Most of these branches just differ from their neighbors in the smallest of ways, by the properties of just one or two particles. When the branches are so similar, we can usually lump them together when calculating probabilities for observational outcomes—distinctions without a difference having no effect.

We are back to a kind of probability familiar to us from our description of thermodynamics, probability needed only to describe our ignorance of the fine details of the system we wish to describe. In the many-worlds picture, there would be no need for probability if we could look at the wave function as a whole, just its deterministic evolution over time according to Schrödinger's equation: if we knew how the wave function started—the birth of the multiverse—we could calculate it for all time.

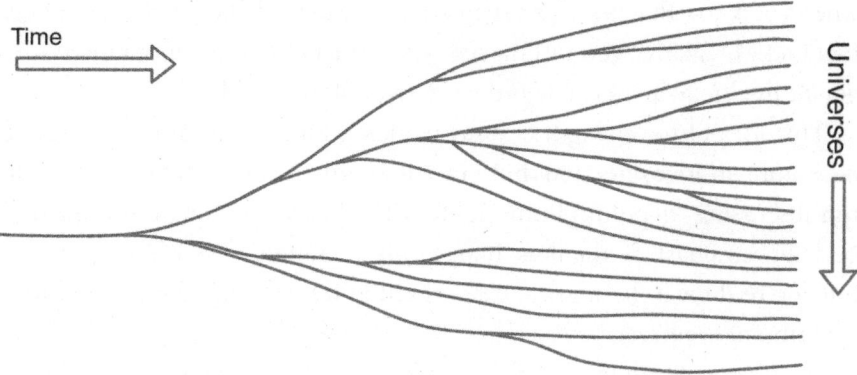

Figure 14. The branching multiverse. Each line represents a single universe.

But we don't have that view from above and outside, just as we don't have a list of all the locations and speeds of all the particles in a gas. All we have is the branch of the wave function in which our different copies exist and the probabilities that we can use to determine which branches we might be in.

So, the phrase "many worlds" is in some ways the most parsimonious of interpretations of the mathematics of quantum mechanics. Its adherents sometimes call it "wave function fundamentalism" and would say that it takes the mathematics very seriously, alongside reifying the wave function itself as an actual object that encodes everything that there is to know about both the contents of their branch of the entire multiverse of which the branches are a part.

The cost of getting rid of measurement and collapse is the enormity of the wave function and the plethora of worlds that it describes. The concept of many worlds is theoretically parsimonious but "ontologically extravagant":[16] everything, everywhere, all at once, encoded in the wave function and its evolution. This extravagance is the first objection that many have to this interpretation of quantum mechanics.

In a 1971 short story, "All the Myriad Ways," Larry Niven shows us a world in which the many-worlds interpretation is known to be true — scientists have worked out how to travel between branches — and the repercussions are devastating.[17] Life is pointless, the inhabitants of this reality find, when you know that there's a version of you who made better choices, or got that lucky break that you just missed—there's a doppelgänger out there who got *all* the lucky breaks. The outcome is a rash of suicides.

This idea of traversing the many worlds is taken in what is intended to be a more positive direction (but I'm not so sure) by philosophers and scientists discussing "quantum immortality" (its other name, "quantum suicide," perhaps emphasizes the less happy interpretation):[18] Schrödinger's cat, but you're the cat. It's a grisly experiment to test whether the many-worlds experiment is correct.

Hook yourself up to a machine with some sort of quantum interaction. Depending on the outcome, you might live on, or be sedated and euthanized, with a 50 percent chance of each outcome occurring. But in some branch, there's a version of you that lives on. Now, repeat the experiment, again and again. Each time, the chance that you survive is 50 percent, and this accumulates after each step: 1/2 after the first step, then 1/4, then 1/8, 1/16, 1/32, and on and on by powers of two. In most branches, you're dead. But that one live branch is crucial, because that's the only one where your consciousness resides.

In other interpretations of quantum mechanics where there is a single Universe, it is highly unlikely that you would still be alive after, say, 10 steps—just one chance in 1,024. But if there are many worlds, it is *certain* that there is a copy of you who survives (too bad about all the versions of you who didn't). Using Bayes' theorem, this is just the likelihood: in the many-worlds picture, the probability that some version of you finds yourself alive after many steps is just one—with certainty. But under all other interpretations, that probability reduces by those powers of two: 1 in 1,024 after 10 steps, and more like 1 in a million after 20 steps. So if you're still around to even have an opinion, you should probably assume that the many-worlds interpretation does, in fact, hold. I don't think any of my colleagues in experimental physics would be willing to go through with this experiment (and I don't think it would pass the health and safety checks).

If it does hold, the many-worlds interpretation must still contend with the technical implications of Bell's theorem. Whenever any quantum mechanical interaction occurs, branching the wave function, a new world—a new universe—is created: a whole new version of everything that exists, coming into existence instantaneously. This is nonlocality, as ensured by Bell's theorem. Mathematically, this may not be mysterious, but physically and ontologically it is hard to grasp, and leads to the most pressing puzzle of this way of looking at quantum mechanics: How and why do these branches—just terms in an equation that we write down—become worlds, physically realized and existing universes?

The theory itself does not have much of an answer to this question, only that it relies on its definition that identifies worlds with branches of the wave function.

Other details of the branching process and the worlds that it creates are also not completely specified by the many-worlds idea alone. For any individual quantum mechanical interaction, there are many ways we could write the wave function as a sum of branches, but most of them do not look like a world with a single outcome, with Alice seeing a particle spinning up or down, and Bob seeing the opposite.[19]

One possible solution takes advantage of the ubiquity of these world-creating interactions. Even if Alice and Bob are performing a joint quantum mechanical experiment in their far-apart labs, concentrating on a single pair of electrons, the world around them is constantly undergoing myriad other interactions out of their control and not directly observable.

This might sound like a disaster for the many-worlds idea. Instead of just two branches, one where Alice sees her electron spinning up and another where she sees it spinning down, now there are branches for every possible outcome of all the unobserved interactions.

But the proponents of the many-worlds interpretation see this potential problem as an opportunity. These interactions are themselves *local*: they happen at a particular place at a particular time. Taken together, this has the effect of grouping the branches together: one group with Alice observing the particle spinning up, and another with her observing it spinning down. This process, known as decoherence,[20] comes out of the basic mathematics of quantum mechanics, and in the many-worlds interpretation of quantum mechanics it is responsible for the existence of branches that look like the world (or worlds) in which we live. Critics of the many-worlds interpretation say that decoherence—which applies in any way of looking at quantum theory—isn't enough.[21] The grouping of the branches, which can only be approximate, is not enough to pick out worlds that look like the one (or ones) that we observe.

QBism

The final interpretation that I will discuss in some detail—and in many ways the one that I subscribe to when I'm not just shutting up and calculating—is called QBism. Rather than postulate the continued existence of branches and worlds, QBism takes the mathematics of quantum mechanics seriously. Whereas the many-worlds interpretation starts from the primacy—and then the reality—of the wave function, QBism takes as its starting point the Born rule, alongside de Finetti's slogan "probability does not exist." Quantum mechanics is a tool for adjusting our probabilities in the face of a specific kind of model. These probabilities are Bayesian because, especially to a QBist, there are no other kinds of probabilities. (The name "QBism" started as a shorthand for "Quantum Bayesianism" and has emerged from a group of researchers around Chris Fuchs at the University of Massachusetts and Ruediger Schack at Royal Holloway University in London.)[22]

The wave function just encodes our judgments about a system, and QBists have shown that in many, or perhaps all, cases we can think of, the wave function is a different way of writing down our judgments about the results we would expect if we performed enough measurements on a system to completely specify its state.[23] These probabilities fully encode the wave function. The Born rule, written in terms of these probabilities, gives us the different probabilities that we should use to predict the outcome of any *other* experiment that we (or someone else) might perform. This rule ends up looking similar—but not identical—to Bayes' theorem.

And we should not expect it to look exactly like Bayes' theorem: for example, I have the probabilities that I would assign to the outcome of Xavier's experiment X, but I want to use them to calculate the probabilities for Yvonne's different experiment Y.

Let's return to our slogan: all probabilities are conditional. So I know the probabilities conditional upon performing X experiment, and quantum mechanics tells us the probabilities conditional upon instead performing Y experiment.

This is not so different from physics without quantum mechanics: if I assign probabilities to, for example, the height from which Xavier is going to see me drop a ball, Newton's law of gravity and his equations of motion let me transform these into the probabilities for the experiment of measuring the time at which Yvonne will detect the ball hitting the ground.

We would not expect to use Bayes' theorem here, because the relation between the probabilities for experiment X and experiment Y is a law of physics, not a law of probability. It happens to be particularly simple in the case of classical mechanics, since every possible outcome of Xavier's experiment maps onto a specific outcome for Yvonne.

The quantum situation is more complicated: specific outcomes for Xavier's experiment no longer correspond to specific outcomes for Yvonne. This is exactly the Born rule in action: the laws of physics themselves are only probabilistic. But probabilities—following our other slogan, courtesy of de Finetti—do not exist.

For a pair of classical experiments, there are no physical restrictions on what probabilities I can assign. I need to obey the laws of probability, so they must be numbers that are greater than or equal to zero, and add up to one, but that's it. Some probability assignments may be more reasonable than others—if I know that the experiment is to be performed in a laboratory, I should probably assign zero probability to Xavier dropping the ball from a mile up, or from Mars or the Andromeda Galaxy. Those conditions would be silly, but I could still coherently translate the probability of them into expectations for Yvonne.

The quantum world, however, brings restrictions on what probabilities a user of the theory can assign. I can make assignments that are not just silly, but force the mathematics to fail—not all possible assignments of probabilities correspond to wave functions.[24]

QBist researchers are still working out the repercussions of the mathematical analogy between the Born rule and Bayes' theorem, and what it can tell us about the quantum world. Just as the laws of classical physics connect the probabilities of two different measurements, so do the laws of quantum mechanics. The difference is that, in the QBist view, the quantum laws are only about probabilities—they describe how we should think about the world, but not about the world itself, its "ontology."

This rigorous insistence that probabilities, even in quantum mechanics, are only judgments about the world allows QBists to escape some of the supposedly unpalatable repercussions of Bell's theorem. The wave function may be nonlocal, but it only encodes my probabilities, my judgments about the Universe, and there is no restriction on the kinds of facts about which I should be allowed to have judgments. In the EPR experiment, if Alice sees her particle spin up, she becomes certain that Bob will see his spin down, but this judgment is completely local to Alice—it's in her brain, written down in her notebook, and it can even be transmitted to Bob, but only by local interactions that do not violate any laws of physics (telephone, email, carrier pigeon, and so on). For a Bayesian this change is no different than any other change in my knowledge of the world. When I get more information, I might assign different probabilities to a particular experiment.

This point does force the question of the QBist ontology. Quantum mechanics gives us the Born rule, a new way to manipulate probabilities, but where does this rule come from? If the wave function does not exist (except as a tool in our heads), what *does* exist? What is it about the world that makes it quantum? How do we reconcile a world of stuff, about which we seem to be able to say objectively true things, with a theory whose only use is to manipulate (our) probabilities? The QBists don't yet have a complete answer to this question, but like most of the most interesting parts of physics, that incompleteness is at least part of what makes it an exciting research project.

All the Rest

There are many other interpretations of quantum mechanics, and I am going to apologize to my colleagues who may be partisans of one of the camps that I am not discussing here in depth, or at all.

Other ways of looking at quantum mechanics depart from the Copenhagen interpretation in more significant ways. Some say that quantum mechanics, as we learn it from our textbooks, is not quite the correct theory. They start from the feeling that a physical theory of the Universe should not be inherently about probabilities. Rather, the theories should specify everything that there is, even if we cannot know it.

These theories try to rewrite the laws of quantum mechanics in the same way, so that probabilities arise only because of our ignorance of more fundamental quantities that we cannot directly measure. But those "more fundamental quantities"—hidden variables—still must obey the constraints imposed by Bell's theorems.

Louis de Broglie, one of the physicists involved in quantum mechanics during its earliest years, saw how the wave function could be recast, along with some additional mathematics, to make a theory in which the properties of particles are completely determined (the theory was independently rediscovered by David Bohm in the 1950s) by the complex wave function, which gets split into a part describing the probabilities and the *pilot wave* from which the theory gets one of its names. It is a "hidden variable" theory in which the pilot wave pushes particles around. The particles' positions and speeds appear to be random, with probabilities appearing because of our ignorance, but obeying the equations of quantum mechanics—namely Schrödinger's and Born's. The pilot wave itself acts and reacts faster than light, so Alice and Bob see their expected results and obey Bell's theorem. (The de Broglie–Bohm theory inspired Bell to work out his theorem in response to the EPR paradox.)

An idea called *superdeterminism* stipulates that the details of Alice's and Bob's separate experiments are linked, despite Alice and Bob being too far apart to communicate directly. I've alluded to the possibility that one way to set up this long-distant link is to disallow free will: in such a clockwork Universe, quantum mechanics seems probabilistic and random, but in fact all choices have been prescribed since the beginning of time. This realization of quantum mechanics seems conspiratorial—how did the initial conditions of the Big Bang, 14 billion years ago, somehow encompass the fact that Alice and Bob would be performing their experiments today? Even so, why should the Born rule for probabilities even seem to apply? Considering the puzzles that come from this idea, recent versions of superdeterminism are more nuanced.[25] They add a superstructure on top of the usual quantum mechanical wave function and laws, disallowing those outcomes that don't obey the known laws, always giving the same results as ordinary quantum mechanics. Alice and Bob—and their experiments—are still independent of each other, but the laws of physics are subtly changed so that Bell's equations take exactly the right form. Actually, not *exactly*: there are small deviations from the quantum mechanical predictions, alas so small that they may be unobservable in practice. (As we will see, however, one generation's unobservably small effect can soon become experimentally accessible to a later generation with the progress of technology!)

Another way to solve the mystery of quantum mechanical measurement is to use the distinction between the quantum and classical worlds as a feature, not a bug. In *objective collapse* theories, the wave function is real, but there is no special moment of measurement causing collapse; rather, collapse happens spontaneously and randomly, more and more likely for larger and larger objects. In one version of these ideas, developed by the Nobel Prize–winning scientist Roger Penrose, gravity is responsible for the spontaneous collapse of the wave function. In that version, Schrödinger's equation is just an approximation, with modifications forcing the particles to their eventual locations. Collapse is now determined by these modifications, rather than a moment of measurement combined with the Born rule. But the collapse is still nonlocal, still obeying Bell's theorem.

Most of the time, I am content to just "shut up and calculate," essentially using the tools and language of the Copenhagen interpretation—wave functions that evolve because of the Schrödinger equation and collapse upon measurement. And when I've had to teach quantum mechanics to my undergraduate students, this is the way that I've described it, enough to work out the numbers my students and I need.

By design, all these interpretations and reformulations make the same predictions as ordinary quantum mechanics, with at most tiny, and so far unobserved, differences in some cases. Since I am a Bayesian, to me this means that they all have the same likelihood.[26]

So we are left with our priors to decide which interpretation we prefer. When I have the time and inclination to think about the problem more deeply, I fall back to those interpretations in which my Bayesian view makes the most sense (to me). In the many-worlds interpretation, probability arises from our ignorance about the branching structure of the multiverse, leaving us uncertain about which branch we find ourselves in. And in QBism, Bayesian probability is baked in from the start, allowing—requiring—our judgment about every aspect of the world.

Both the many-worlds interpretation and QBism are not yet fully formed theories of the world; they are more like Lakatos's "research programs." The many-worlds interpretation is very clear that the wave function is fundamental but still needs to account for the rest of its ontology: How do the branches of the wave function become worlds? And it still needs to account for the details of probability—why, exactly, does the Born rule apply?

QBism, on the other hand, is very clear about probability, which pretty much defines the interpretation. Probabilities are personal judgments, and nearly everything in quantum mechanics must be interpreted as the judgment of a user of quantum mechanics. This makes QBism vulnerable to charges of solipsism—"me, me, me," as Chris Fuchs himself has put it,[27] and the lack of an ontology—if everything is a judgment, then what is real?

Though more than a century old, quantum mechanics still feels like a new way of looking at the world, not completely understood. Still described by unbroken physical laws, but also still subject to an uncontrolled, underlying randomness. Bringing back some of our terminology to describe it, we could say it is *naturalism*, but not *determinism*.

In most interpretations, quantum mechanics is a source of uncertainty, perhaps the only fundamental uncertainty in the Universe, due not to our ignorance but something about the makeup of the world.

Most days, I am reassured by this idea that the future is unwritten, that our interventions in the world make a difference—although I also wish for a little more clarity about what it all means.

PART THREE

Cosmology in a Random Universe

TEN

The Oldest Light

I'm a cosmologist—I study the history and evolution of the Universe. These are subjects that humans have explored for as long as we have records of our thoughts. For most of that time, the models we created were largely unconstrained by data: the world is a big, dangerous place; change is slow; humans are small, smart, and sometimes as dangerous as anything else in the world. And humans turn out to be really good at inventing and telling stories (that is, at building models). So of course those stories had the multivalent purpose of providing a backing for societies' moral codes and social hierarchies, alongside giving what might be plausible, if unscientific (unnatural), explanations for the world as it was observed.

I was once involved in a public dialogue on "origins" in art, science, and mythology at a museum in Rotterdam.[1] When I asked a scholar of ancient cultures and mythologies about whether the subjects of her study really believed the underlying mythology as if it were fact and not metaphor, she said, more or less, of course not. I now think her deeper point, uncomprehended by me at the time, was that the ancients did not see a sharp demarcation between fact and metaphor.

Perhaps this is unsurprising: the Universe is vast, and we can observe only an infinitesimal portion of it. Before the scientific revolution, before Newton's understanding that the same laws that apply here on Earth would apply everywhere, it would not have occurred to many people to even try to find a scientific explanation for the origins of the Universe.

After Newton presented his laws, it became possible to build a physical model for the Universe and our not particularly special place in it. Scientists had begun to take seriously the possibility that the sun was one of many stars, an idea that goes back at least to ancient Greek philosophers like Anaxagoras. In the 1660s, the Scottish natural philosopher James Gregory used this model of the sun as a star to work out the distance of our solar system to those more distant stars. Objects look brighter as they move closer, precisely as the square of the distance between them. If all we know is that all stars are approximately the same intrinsic brightness (astronomers use the term "luminosity"), then the relative brightness of a star to that of the sun gives us the square of the distance of the star relative to that of the sun. Stars appear much dimmer than the sun—and hence must be much farther away. Exactly how much farther is a matter of experiment that carefully compares the brightnesses of the sun and stars. Gregory did this by using the reflection of sunlight off Jupiter, which could be observed at night, and using the fact that he could wait until Jupiter was at the right distance to match the brightness of a star. In 1668, this method enabled Gregory to estimate the apparent brightness of the sun to be about seven billion times that of the star Sirius. This is within about 25 percent of the correct answer for the brightness, although Gregory's distance is much more discrepant, since another part of the model he used is wrong: Sirius is *not* the same luminosity (intrinsic brightness) as the sun: it is about 25 times more luminous and hence about five times further away than predicted by this analysis. (Christian Huygens, one of Newton's rivals in the quest for the law of gravity, did a similar experiment, using only his memory to compare the sun and Sirius, and found that the ratio of their brightnesses to be about 750 million, a factor of about ten below Gregory's result.)

But even these not-quite-correct calculations started to give some idea of the scales required to create a model of the Universe. The hundreds of millions of miles needed to describe distances within the solar system are tiny compared to the distances between stars.

In the eighteenth century, this model was taken yet further. On a dark night, the pattern of stars is dominated by the Milky Way stretching across the sky. Our word "galaxy" comes via old French, mediaeval Latin, and originally the Greek γαλαξίας ("galaxias"), also meaning milky. Nowadays, the term "Milky Way" refers exclusively to our own agglomeration of hundreds of billions of stars, but "galaxy" applies to any such system of stars. Thomas Wright, an astronomer in Durham, England, saw that this model places the solar system within a disk-like agglomeration of stars.[2] This idea was taken seriously by philosophers like Immanuel Kant in the mid-eighteenth century, who hypothesized that our Milky Way might be just one of many "Island Universes" within a much vaster space.[3] (Kant was invested yet further in astronomy: along with Laplace—whom we met due to his work on probability—he also put forward the idea that the solar system could have condensed out of a rotating ball of gas and dust. This is the so-called nebular hypothesis that is still at the center of our current theories of star and planet formation.)

In the late eighteenth century, the French astronomer Charles Messier compiled a catalogue of about one hundred bright fuzzy patches visible with telescopes at the time, distinct from both planets and point-like stars— his designations are still used today when discussing these objects in the astronomical literature. Some were clearly agglomerations of hundreds or thousands of stars, some seemed to be nebulae of glowing gas, but many could not be identified. Were they more like the nearby diffuse nebulae of gas and dust, or were they much more distant island universes of their own?

The distribution of stars within the Milky Way was further investigated by William and Caroline Herschel, who used their observations to make a map of the Milky Way (Figure 15). Visually, our solar system seems to be near the center of the disk, although we now know that we are on the outskirts— but we can't see through the center across to the other side because of the presence of small particles of dust threaded through the disk, blocking our view of all but relatively nearby stars.[4]

The full shape and extent of the Milky Way, and the sun's location within it, was not appreciated until well into the twentieth century. A subset of Messier's objects, the so-called spiral nebulae, were singled out as candidates for island universes outside of the Milky Way, but there was no consensus.

In 1920, the split between those who thought all of the nebulae were within our Milky Way and those who thought that some were vast galaxies of their own was so pronounced that the US National Academy of Sciences

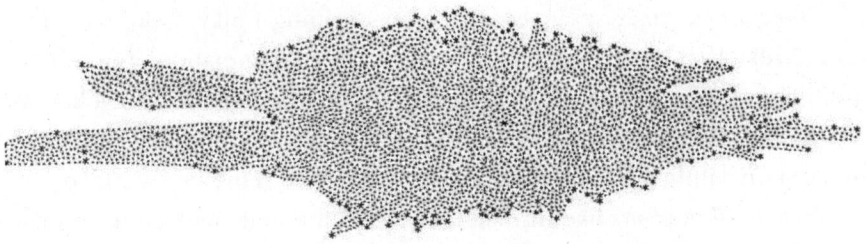

Figure 15. The Milky Way as drawn by Herschel from observations of nearby stars he made with his wife, Caroline Herschel. ("On the Construction of the Heavens. By William Herschel, Esq. FRS." *Philosophical Transactions of the Royal Society of London* 75 [1785].)

decided to host what has become known—with no sense of exaggeration within the astronomical community—as "the Great Debate" between Harlow Shapley and Heber Curtis.[5] Curtis was an astronomer at the Lick Observatory (affiliated with the University of California) and defended the Kantian idea that many of the nebulae were separate island universes. Shapley worked at Mount Wilson, a telescope observatory in Pasadena, near Los Angeles, elsewhere in California (then one of the best sites for astronomy in North America, despite the proximity of what was becoming one of the brightest and largest American cities). At the time, there was no dispositive evidence one way or the other. Shapley had several main arguments. First, at least some of Messier's objects clearly were within the Milky Way, as determined by the brightness of some of the individual stars within them. Second, if some of these nebulae were outside of the Milky Way, but with a comparable number of stars to our own galaxy, it would imply an immensely vast overall scale for the Universe, mostly filled with empty space. This was not disallowed by the physics, but seemed incomprehensible to many astronomers. Finally, he did have some seemingly hard data on his side: the Dutch American astronomer Adriaan van Maanen had purported to have observed the rotation of one of these spiral nebulae, M101, now also known as the pinwheel galaxy and shown in Figure 16 (I've given the game away with the name, but no one will be surprised by the spoiler).[6] Van Maanen measured the apparent rotation speed on the sky, corresponding to a rotation around the whole galaxy every 100,000 years. At the distances necessary

Figure 16. Messier object M101, the pinwheel galaxy. (NASA, ESA, CXC, SSC, and STScl.)

for the island-universe hypothesis, however, this would translate into speeds greater than the speed of light, disallowed by physics and common sense.

Curtis's arguments in favor of the island-universe idea were also circumstantial. The Andromeda nebula (M31; now known as the Andromeda Galaxy, of course) had almost as great a number of easily observed flaring stars called novae as the entire Milky Way—why would a nebula within our galaxy house quite so many in such a small volume? Moreover, their brightnesses were such that they would be much farther away, unless they were instead a very dim, otherwise unknown population of nova-like objects. Moreover, the structure of these spirals was remarkably similar to what had been learned about the Milky Way: they were in disks, dominated by dark clouds of dust—the same as what obscures our view of more distant stars. And there was even beginning to be evidence that the Milky Way had a spiral pattern of its own. Curtis did admit, however, that van Maanen's measurements of the rotation of M101 would disprove his thesis.

From the same data, Shapley and Curtis had come to very different conclusions—by using different models.

But van Maanen was wrong, and so was Shapley, and so, therefore, was Shapley's model. The proof was to come from Shapley's home institution, Mount Wilson Observatory, with the work of Edwin Hubble starting in 1923 (although Shapley had moved to Harvard to take over as director of the observatory there in 1921). Hubble had discovered another special kind of star in Andromeda: a Cepheid variable, a star that pulsates at a regular period. About 15 years earlier, Henrietta Swan Leavitt, a Harvard astronomer, found that these stars had a strong mathematical relationship between the period of their pulsation and their luminosity (intrinsic brightness). Leavitt had discovered the first "standard candle" with a known luminosity, and it then became possible to revisit the calculations of Gregory and Huygens and properly work out the distances to at least some stars from our solar system. Cepheids continue to be used to measure the scale of the Universe, a practice started by Leavitt and mastered by Hubble: observing Cepheids in Andromeda, he could show that Andromeda was, indeed, much farther away than any of the stars known to be within the Milky Way. Stars within the Milky Way Galaxy are millions of times farther apart than planets within our solar system; galaxies are, in turn, millions of times farther apart than the stars within a single galaxy. Because of this, it doesn't really matter whether we measure distances to these galaxies from the Earth, or the sun, or even anywhere else within the Milky Way.

Hubble continued searching for Cepheids in what were now known to be external galaxies, and in 1929, he combined the distances to these galaxies with measurements of their speeds. These measurements were pioneered by Vesto Slipher in Arizona and Hubble's assistant, Milton Humason, who had worked his way up from being part of the construction team of Mount Wilson Observatory and earned his doctorate in 1957.[7] They determined the speeds of the nebulae containing the Cepheids using the Doppler effect, with the colors of the stars shifting toward blue or red if the stars (and their host galaxies) were moving toward or away from our galaxy. (The full calculation is a little more complicated than just looking at the overall color of a star: we measure the Doppler shift by observing at individual lines in the star's spectrum, produced by the presence of specific atoms in the star's atmosphere, just as had been first seen in the sun's spectrum in the nineteenth century, and that had just been explained by the discovery of quantum mechanics in the previous few decades.)

At around the same time, Georges Lemaître, a Belgian astrophysicist (and Catholic priest), was one of the few experts working out the cosmological repercussions of Einstein's gravitational theory, general relativity, and saw that an expanding Universe was a natural possibility. He too was aware of both the Cepheid and Doppler shift observations, and saw that they might be evidence for his model.

Comparing the measurements of distances and speeds, Hubble and Lemaître independently found that they were related.[8] Almost all galaxies were moving away from us, and more distant galaxies were moving away faster, the speed exactly proportional to the distance: a galaxy twice as distant was moving away twice as fast. Nowadays, we write this model as

$$v = H_0 \times d.$$

In this equation, v is the speed (velocity) of the galaxy, d is its distance from us, and H_0 is called the Hubble constant. The equation is called, naturally, the Hubble-Lemaître law (although it went by the name "Hubble's law" until 2018).[9]

The observed Hubble-Lemaître law only involves quantities measured from Earth, so it directly describes distances and velocities relative to our position in the Milky Way. But its implications are universal: the Universe is expanding uniformly in all directions.

Imagine that there were galaxies located at the intersections of the grid shown at the top of Figure 17. If we stretch the grid, keeping it square so it appears like the grid at the bottom of the figure, then every single galaxy sees all the points (other galaxies) around it moving away, the same in all directions. Every point sees its own equivalent Hubble-Lemaître law. Copernicus's revolution, displacing Earth from the center of the cosmos, continues: there is no center, and we — humans, Earth, the sun, the Milky Way — do not occupy a particularly special place.[10]

Lemaître himself, along with Alexander Friedmann, Howard P. Robertson, and Arthur Geoffrey Walker (a group often referred to as FLRW) saw

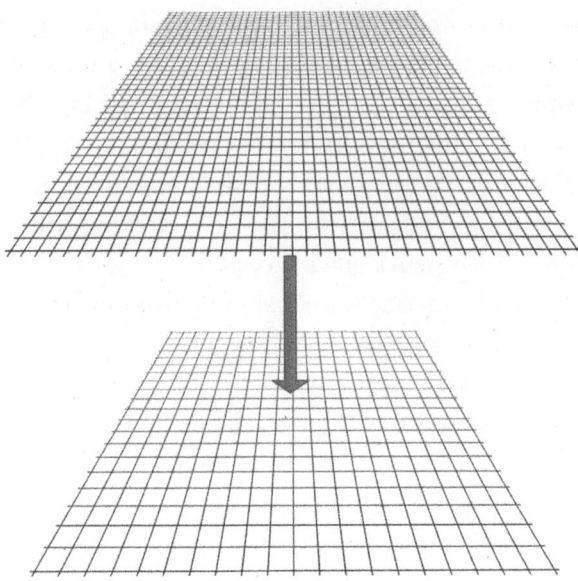

Figure 17. A two-dimensional representation of the expanding Universe. Everything is moving further apart from everything else, but there isn't a center.

this expansion as a consequence of Einstein's theory of general relativity, which allowed them to trace this expansion forward and backward in time. Current measurements imply that the Universe, or at least the part of it that we can observe, has been expanding for about 14 billion years, and will continue to expand, ever faster, perhaps forever. This doesn't mean, necessarily, that everything was created 14 billion years ago; rather, we can only say that our part of the Universe started its expansion around that time.

The theory of general relativity also let them see a connection between the expansion of the Universe and the geometry of space-time. Einstein's theory intimately connects gravity with the curvature of the Universe. FLRW showed that the overall geometry of the Universe was a crucial part of the mathematics of their theory, described by a single number. In Einstein's theory, the geometry is determined by the contents of the Universe, summarized by an overall density of energy, now often given the Greek letter Ω (sometimes with a subscript to tell us what density it describes; without such a subscript we'll let it cover the sum of all the stuff in the Universe).

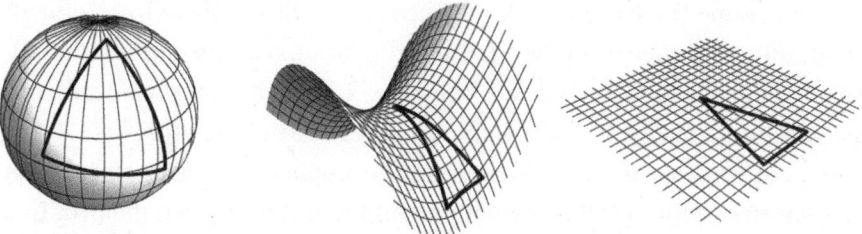

Figure 18. Straight lines behave in a confusing way in a curved Universe (here represented as only a two-dimensional surface). A triangle with the same lengths of its sides will have *different* angles.

We can picture the geometry by analogy (but it's just an analogy) with two-dimensional surfaces, shown in Figure 18. The three simplest possibilities for the shape of a curved surface in three dimensions as well as two are:

- flat ($\Omega = 1$)
- curved like the surface of a sphere ($\Omega > 1$), and
- curved in a saddle shape ($\Omega < 1$).

I've also given the values of Ω corresponding to each of these shapes. This parameter, Ω, is just a number we can measure within the FLRW model, defined to be 1 in the flat case, but the repercussions of its value for the overall geometry of the Universe are so important that we equally well speak of separate models for each of these possibilities, all contained within the FLRW model. In a flat geometry, all the usual rules of Euclid apply: for example, the angles of a triangle add up to 180 degrees, and parallel lines never cross.

On the surface of a sphere, neither of these intuitions are true anymore: a straight line corresponds to a "great circle" like the equator or a line of longitude from the North to the South Pole—these are the lines that feel straight as you travel along them and correspond to the minimum distance between points on the surface. But consider three of those lines making a triangle: if you add up the angles you will always get more than 180 degrees. Similarly, lines that start out parallel, such as those longitude lines, will always cross (at opposite poles in this case). Conversely, a saddle shape also goes against our intuitions: the angles of a triangle sum to less than 180 degrees; parallel lines never stay the same distance apart.

These same three cases, and only these cases, also apply in three dimensions, although they are much harder to picture, and are further complicated by the expansion of the Universe. In practice, we can still use something akin to the triangle property to measure the curvature of our Universe, because that allows us to compare the physical size of an object to its angular size in the sky: the geometry maps the space between the two. In a flat universe, an object's physical size maps to the angular size (in radians) that it would have if observed from some point by a simple geometrical formula: divide the physical size by the distance from the object to the observer. In a spherical universe (technically, a *hyperspherical* universe), the same physical size maps to a larger angular size than in a flat universe; in a saddle-shaped (*hyperboloid*) universe, the same physical size maps to a smaller angular size than in a flat universe.

Since Hubble and Lemaître compared measurements of distances and speeds, the data have grown so constraining that we have incorporated this expansion of the Universe into a detailed model that tells us the history of the Universe over the last 14 billion years. We've been able to develop that model despite only observing the sliver of the Universe that happened to have something shining light exactly in our direction at just the right time for our eyes, or our telescopes, to see it.

So most of the light in the Universe never makes its way to Earth, but from this tiny fraction of all the light in the Universe we have created a model that accounts for its history that stretches back to its origins.

Over those 14 billion years (remember that a billion is 1,000,000,000, or a thousand million, or a thousand thousand thousand) everything that we can see has been expanding from an incredibly dense and compact region. This idea is called, of course, the Big Bang, and it can be really hard to get your head around it (even if you're a professional cosmologist).

But if the Universe is expanding now, what does that say about the past? How did we get here? If things are moving apart as we look into the future, then earlier things must have been closer together. The further back we go, things were closer and closer together, making the Universe denser and denser. What we have learned about thermodynamics then lets us say even more about the early Universe: we also know that things get hotter and hotter as we compress them.[11] So the early Universe would have been both hotter and denser than it is today. This is what we call "the hot Big Bang model," which says the Universe has been cooling and expanding over the eons until today and will continue to do so into the future.

So the baby Universe was hot and dense, and as it enters its teenage (14 billion) years it is very different than it once was. And just as babies change much more rapidly than teenagers or adults, so was the very early Universe a time of transformation.

In fact, we are pretty sure that the Universe didn't only start *dense* and *hot* but also *simple*. Nowadays, there are galaxies and clusters of galaxies, stars and planets, and, at least here on Earth, animals and people, trees and buildings, television sets and computers. One way we can measure that complexity is by comparing different parts of the Universe to one another. The conditions on the surface of Earth are very different from those in interstellar space and also very different from somewhere inside a star—there is a lot of variation. But if we go back far enough, well before the first stars and galaxies, we find that the Universe was much more uniform.

How do we know this? We know it by using the same combination of model and data that I hope we've come to expect by now. And that combination tells us much more about the history of the Universe.

Ever since the discovery by Hubble (aided by Humason, and Leavitt, and Lemaître, and many others) of the expansion of the Universe, cosmologists have been trying to incorporate this knowledge with our understanding of gravity, thermodynamics, and quantum mechanics to build a coherent model of, well, everything: a cosmology.

Starting from the idea that the Universe is expanding and cooling over time, we can trace things backward, thinking about the different physical processes that come into play as the temperature—energy—increases into the past.

Why Is the Universe Interesting?

The best way to understand the different events that have marked the history of the Universe on the largest scale was first written down by Andrei Sakharov. Sakharov was known as the father of the Soviet Union's hydrogen bomb and, decades later, as one of its most famous dissidents, for which he won the Nobel Peace Prize in 1975. But he also found time to think about the history of the Universe.[12]

He was specifically worried about why the Universe seems to be made predominantly of what we call matter, rather than antimatter. For every particle of matter—electrons, protons, neutrons—there is a matching antiparticle—anti-electron, antiproton, antineutron (although the anti-electron is more commonly known as the positron). When a particle encounters its antiparticle, the pair can easily be converted into a pair of photons, in a process known as annihilation. We know that the Universe is primarily matter, because we see no evidence of these annihilation processes that would have to occur if there were regions of antimatter bordering regions of matter, releasing light from these annihilation processes, observable as a background of high-energy photons, X-rays and gamma rays, throughout the Universe.

Today, therefore, particles vastly outnumber antiparticles. Antimatter only seems to be made in high-energy collisions in particle accelerators and extreme astrophysical environments. How can our model explain why the Universe seems to be made mostly of matter (a word we use because it's what we are made of, and hence was named first)?

Sakharov realized that there are three conditions that need to occur to end up with any excess of matter over antimatter:

1. It needs to be possible for interactions that start with equal amounts of matter and antimatter and end up with more matter than antimatter to occur.
2. If those matter-creating interactions do happen, it cannot also be the case that they are exactly mirrored by antimatter-creating interactions.
3. Even if you can make a net excess of matter through the first two conditions, you still need to ensure that the Universe does not immediately unmake the excess by just running the same reactions backward.

All these conditions seem obviously necessary to end up with our matter-filled Universe, but it is not obvious whether—or at least how—the actual

Universe manifests them. The standard model of particle physics does *not* have interactions that obey the first condition; the second condition was discovered to be part of the standard model in the 1960s,[13] but it is still not clear whether this is enough to give the amount of matter that we observe; the third condition is provided by the Big Bang itself, in that as the Universe expands and cools, particles move further apart and slow down relative to one another. This makes interactions less common so the Universe can maintain the excess as it builds up. In the language of thermodynamics, the expanding Universe cannot stay in equilibrium, which requires all possible interactions to have enough time to proceed.

Sakharov's conditions mean that either our model of particle physics needs to be extended to meet his first condition (and possibly the right amount of his second), or, less plausibly, that the Universe just happened to be born that way, with the exact overabundance of matter compared to antimatter that we observe. Cosmologists always prefer to find our solutions in the laws of nature themselves rather than in the happenstance of the initial conditions for the Universe.

Although Sakharov wrote down these conditions specifically for creating an excess of matter over antimatter, they apply more generally to the major changes that occur in the Universe as it expands and cools. Equilibrium is boring, described by a single number, the Universe's temperature. Only by falling out of equilibrium does the Universe build up relics from its past, out of which comes everything that differentiates it from a featureless gas.

The more general version of Sakharov's conditions under which the Universe can evolve from a boring state to a more interesting one is:

1. It must be possible to create the more interesting state.
2. If the more interesting state can be canceled out by its opposite, the two processes must not happen identically.
3. The Universe must be expanding fast enough so that the interesting state is not destroyed as soon as it is created.

Applying this version of Sakharov's insight allows us to build up our Big Bang model: a rough timeline for the history of the Universe and how it became more interesting (out of equilibrium) over time, even if we do not understand the details of all the physical processes.

At the start of its expansion, the Universe was an incredibly dense, hot, featureless gas. Thermodynamics tells us that the temperature of an object is directly related to the average speed of the particles within it, and at this early time—less than 10^{-35} seconds after the Big Bang—this was nearly the speed of light, even for the heaviest kinds of particles.

When particles are moving so quickly in so dense an environment, they lose much of their identity. In a hot primordial gas like the one at the start of the Universe, every possible interaction can occur, and thermodynamic equilibrium is maintained between every different kind of particle. The gas is a homogeneous mix of all kinds of particles: the more familiar quarks, electrons, and photons that make up the visible world, but also their more obscure cousins that, today, we only encounter in particle accelerators, and even some particles that we have never directly observed.

Since just after Hubble's discovery of the expanding Universe, evidence has been building up that there must be particles beyond the ones that make up all the stuff that we see around us. In the 1930s, Fritz Zwicky observed that the total mass of a giant group of galaxies known as the Coma Cluster were not enough to explain the large velocities of the individual galaxies—the cluster should not be able to hold itself together over time. Zwicky theorized that there must be additional mass, not producing light and hence invisible to us, dominating the cluster.

This need for "dark matter," as Zwicky referred to it, has not gone away as our observations have improved. Further cluster measurements show similar behavior, and in the following decades similar behavior was seen within galaxies themselves. In the 1960s and 1970s, Vera Rubin measured the rotation of spiral galaxies and saw a similar phenomenon—the large speeds could only be explained by much more mass than could be accounted for in the stars, gas, and dust she could directly observe.

Dark matter remains central to the Big Bang model, but its details are scarcely better understood today than they were when Zwicky coined the phrase or when Rubin confirmed its existence within galaxies. (Alternative theories exist, including modifications of Einstein's theory of gravity, but none of the alternatives have had quite the explanatory power of the elusive dark matter idea—not yet, at least.)

For most of the history of the Universe, dark matter has dominated its expansion, but not the interactions that have left behind the objects that we observe. As the Universe expands and cools, those interactions become less frequent, and the particles' speeds fall well below the speed of light, allowing Sakharov's conditions to be met. Early on, this meant that a small excess of particles over antiparticles was built up, only about a billion and one particles for every billion antiparticles.

The next major event in our timeline of the history of the Universe happened at about a microsecond after the start of the expansion. At this point, the Universe had cooled enough that the quarks came to be confined in pairs and triplets. Two sorts of quarks, named "up" and "down," are the building blocks of the neutrons and protons that themselves make up the nuclei of atoms, but there are two other pairs ("charm" and "strange"; "top" and "bottom") of increasing mass. The triplets make up the neutrons and protons and their heavier, less stable cousins, all called "baryons" (after the Greek βαρύς—barys—for heavy), whereas the pairs make "mesons," particles that help keep the nucleus bound together. After this time, quarks are rarely seen in isolation. But the Universe still had to undergo many changes before it would look like it does today.

After about a second, the Universe cooled enough that almost all matching pairs of particles and antiparticles had annihilated into photons. For every billion such pairs that annihilated, just one matter particle was left behind. This gave us a Universe closer to today's: a billion photons for every matter particle, and almost no antimatter.

For the first minute or so after the start of the expansion, it was still hot enough to split the nuclei of atoms themselves: the matter in the Universe was a soup of individual protons, neutrons, and electrons, along with a bath of photons. As the Universe cooled, nuclei congealed in a process called "Big Bang nucleosynthesis" or BBN.[14]

The details of BBN are very well determined, despite our inability to observe the moment it occurred directly, or to reproduce the same conditions in a laboratory. The model depends on the well-understood physics of atomic nuclei, as well as the state of the Universe at the time it occurred: the rate of expansion, the density of the dark matter, and the amounts of the various constituents determine the speed of the nuclear reactions that create and destroy the lightest elements of the periodic table: hydrogen, helium, lithium, beryllium, and boron. The process was almost complete by about three minutes; from then on, the synthesis of any further elements happens in the only regions of the Universe that get hot and dense enough: the nuclear furnaces in the cores of stars, or the even hotter conditions when some of those stars collapse and explode in supernova explosions brighter than their host galaxies.

The formation of the elements via BBN at three minutes was the last major transition of the invisible Universe, the photons still trapped in the dense and hot plasma of nuclei and electrons, not directly observable by us and our light-collecting telescopes.

Relics of the Early Universe

The Universe was still so uniform 400,000 years later, and so hot, that it was mostly a featureless gas of hydrogen, along with about 25 percent helium and a small admixture of other light elements such as lithium, beryllium, and boron. The dominant hydrogen is made up of a single electron orbiting around a single proton. Electrons have negative electric charge; protons have positive electric charge and are about 2,000 times heavier than electrons. So a single hydrogen atom has the heavy and positive proton at the center, surrounded by the negative electron. The charges cancel out so the atom has no net electric charge. However, when it's too hot, the electrons can't stay in orbit around their protons and instead get stripped off, making what's called an ionized plasma — free-floating positive and negative charges, bouncing around in the extreme heat.

A plasma has one more important property, not at all obvious if you don't know the details of quantum mechanics: light is very easily deflected by the charged particles. So a charged plasma is essentially opaque: light bounces around rather than getting transmitted through it. The light is scattered so much that we just see a uniform glow rather than any details of what might be on the other side. Clouds similarly spread the light of the sun into a uniform gray-white, losing much of the information about their original direction (in this case, light is scattered not because of small, charged particles but rather off the much larger water droplets).

The finite speed of light—exactly one light-year per year—means that the farther away from us that we look, the earlier in the Universe's history we are observing. The Milky Way is a few hundred thousand light-years across, so within our galaxy we are seeing things as they were over the last few hundred thousand years. Beyond the Milky Way, other galaxies can be millions or billions of light-years away, corresponding to us seeing them as they were that many millions or billions of years ago. Eventually, we will get so far away, and hence so early, that the Universe would have been made up of this opaque plasma of electrons and protons.

So there must be a distance, and a corresponding time, at which the Universe went from being opaque to transparent, shown in Figure 19. This picture of the Universe is like the surface of a cloud, beyond which the light just bounces around. But some of that light filters through to the surface of the cloud after which it can move unimpeded on a straight line, and some of that light happens to be pointing toward your eyes (or our telescopes).

Just as we can still learn something about the sun even when we observe it through a cloud, we can learn about the early Universe by observing the light that has scattered off the charged particles present in the early Universe.

As far as we're concerned, there's a glowing curtain back-lighting everything we can see—there's no light directly coming from earlier than this important time when the Universe went from being an opaque, charged plasma of independent electrons and protons to being a transparent, neutral gas of hydrogen. That time of transition, called "last scattering," marks the last time that much of the light in the Universe would have scattered off a charged particle. Nowadays, and nearby, light "scatters" all the time. That is exactly what it means for us to see something: light from the sun or a lamp scatters off some object and into our eyes. But most of the conditions of the Universe aren't like those on Earth. Most of the Universe is nearly empty space, transparent to light. From our vantage point, here and now on the transparent side, we look farther and farther away (earlier and earlier into

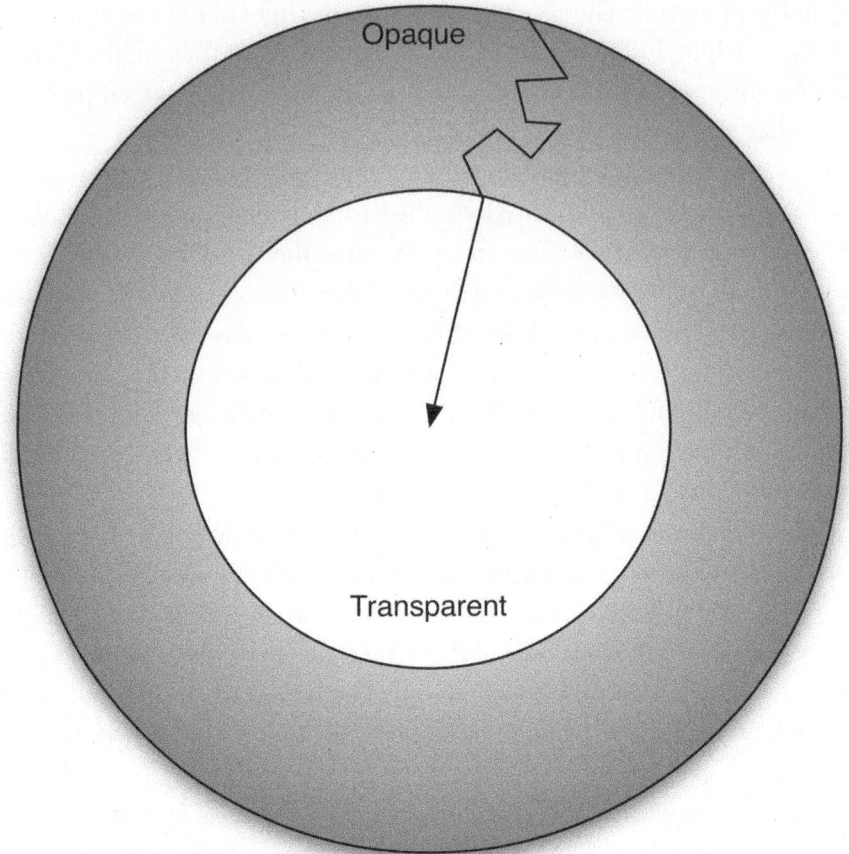

Figure 19. Scattering in the early Universe. We, the observers, are at the center. Because of the finite speed of light, looking farther away is looking back in time, eventually to where (when) the Universe went from being opaque to being transparent. The solid line represents the path of light, scattering often in the early Universe and not at all since then.

the Universe's history) and eventually see light coming from a sphere around us that we call the "last scattering surface."

The earliest successes of quantum mechanics, by Bohr and others, were the predictions of the chemical properties of the hydrogen atoms—the energy of the levels of the electrons orbiting the nucleus. Those same predictions tell us how much energy is required to keep an electron and a proton separate. A little earlier, in what would turn out to be one of the first hints of the quantum nature of matter, Max Planck had worked out how the energy of photons is related to the temperature—faster, more energetic photons correspond to higher temperatures. Putting these ideas together, along with information about the relative numbers of protons, electrons, and photons, we can work out exactly when the Universe expanded and cooled enough for the hydrogen atoms to combine—physicists often call this "*recombination*" even though this was technically the first time the event had happened— from their constituent protons and electrons.

That moment of recombination, giving us the sphere of the last scattering surface, marks, literally, the furthest we can see.

Those leftover photons from the Big Bang are the link between the different transitions that we have learned about in this chapter. The annihilation of the pairs of particles and antiparticles left the Universe suffused in a bath of these photons, a billion of them for every eventual particle of normal matter. That large number of photons enables the physics of BBN to leave precisely the residue of light elements that we observe today—more photons would have delayed the beginning of BBN, keeping neutrons and protons from forming heavier nuclei for longer.

The physics of nucleosynthesis was first understood in a 1948 paper by Ralph Alpher, Hans Bethe, and George Gamow (Bethe was added to the paper by Alpher and Gamow mostly for the joke of making the author list sound like the beginning of the Greek alphabet, $\alpha\beta\gamma$—"alpha, beta, gamma").[15] The dependence on the photon bath, and its repercussions for the Universe today, was discussed in a paper that same year by Alpher and Robert Herman.[16]

The idea that these photons might be observable was mentioned in a 1964 paper by the Soviet physicists Andrei Doroshkevich and Igor Novikov.[17] At nearly the same time, a group at Princeton University in New Jersey, led by the polymath physicist Robert Dicke along with David Wilkinson and Peter Roll, were designing a microwave radio telescope to detect these photons. Elsewhere in New Jersey at the time, the astronomers Arno Penzias and Robert Wilson were working at Bell Labs on a new radio telescope. (This was a time when many of the biggest American corporations—Bell

Labs, named after Alexander Graham Bell, was the original "phone company" and would later become AT&T—had pure research arms with no particular technological agenda.)

Penzias and Wilson were calibrating their newest telescope, but could not eliminate a low-level "hiss" underneath every observation that they made. They didn't quite know what it meant: Was it an electronic artifact from their circuits? Was it an afterglow from nuclear tests? Was it from energetic particles in the solar system? Was it emission from the droppings from a pair of pigeons that had started to roost on the antenna?[18] No, no, no, and no. Penzias and Wilson had data, but no model to interpret it (yet); down the road at Princeton, Dicke's group had a model, but no data (yet). Eventually the groups were put in contact, and they all realized what they had seen.[19]

The results were published as a pair of papers in the *Astrophysical Journal* in 1965. The first described the cosmological implications of the data (along with the beginning of the Princeton team's efforts to observe it themselves); the second, less than two pages long, described just the experimental results. The papers referenced each other, and the only mention in Penzias and Wilson's experimental paper of cosmology or any underlying astrophysical or cosmological explanation for their results is this statement: "A possible interpretation for the observed excess noise temperature is the one given by Dicke, Peebles, Roll and Wilkinson (1965) in a companion letter in this issue."[20]

The implications from this pair of articles was big enough news to make the front page of the *New York Times*. The headline "Signals Imply a 'Big Bang' Universe" appeared above the newspaper's fold in company with separate articles about future New York City mayors Abe Beam and John Lindsay.[21] Penzias and Wilson went on to win the Nobel Prize in 1978. Jim Peebles (specifically cited in the paper by Dicke's group as coming up with the details of the cosmological interpretation), one of the great theoretical cosmologists of the late twentieth century, won the Nobel Prize in 2019.

We now call this signal, the bath of leftover photons from the Big Bang, the cosmic microwave background, or CMB. Penzias and Wilson's observation was only sensitive enough to detect the overall intensity of the CMB averaged over the sky. We can describe that intensity as a temperature, which they measured to be about 3 Kelvin (degrees above absolute zero—a very cold temperature resulting from the long period of cooling since the hot Big Bang), but with the Princeton group's interpretation it was already understood that it would not be constant over the whole sky.

For the Universe to be filled with the stars, galaxies, and clusters of galaxies that we observe today, the seeds of those structures had to be present in the early Universe: small fluctuations that would grow by their own gravitational attraction over time. Over the course of the 1960s and 1970s, theoretical cosmologists worked out how those structures would grow and how that growth would affect the photons of the CMB.

The first variations in temperature were seen in 1969 and confirmed through the 1970s—the CMB is about 0.1 percent brighter (about 3/1000 of a degree) in one direction than its opposite.[22] This asymmetry is due to the motion of the Milky Way through the cosmos: a small change in temperature is exactly equivalent to a change in the frequency of the light. Just as the light of distant stars or galaxies is Doppler shifted due to their motion away from us, we see the CMB blueshifted in the direction toward which we are moving and redshifted in the antipodal direction from which we are coming.

This pattern from our velocities can be predicted directly from Einstein's theories and stretches over the whole sky. On smaller scales, fluctuations in the temperature also arise, now measured in millionths of a degree. These trace the small departures from smoothness that will, as the Universe evolves, grow into galaxies, and even larger clusters of galaxies, and all that they contain. A model for the growth of those structures over the history of the Universe also comes from Einstein's theory of gravity and implies that the seeds for today's galaxies would already be in place 400,000 years after the start of the Universe's expansion, departing from smoothness at that time by only a measurement at a level of a few parts in a hundred thousand.

These temperature differences were so small that they had not been observed by attempts to detect them with the technology of the 1970s and 1980s. Cosmologists were beginning to worry that they might not be seen and speculated that perhaps the seeds for today's galaxies had not yet been planted even 400,000 years after the Big Bang, but instead somewhat later, in a process dubbed a "late-time phase transition." Phase transitions are phenomena of thermodynamics like the freezing of ice, which can result in imperfections such as cracks in the ice arising out of an initially uniform medium. In a cosmological context, this transition would not be between liquid and solid, but one that could still move enough matter around to, possibly, leave trace fluctuations large enough to grow into the large-scale ones observed today.

As a graduate student I tried to tease out the repercussions of such a theory. I had arrived at graduate school in 1989 and started working with my supervisor, Professor Josh Frieman, in 1990. We started on this, my second paper with Josh, in 1991. Our main conclusion was that building up structure in the Universe only after the CMB was released from the primordial plasma would usually give *larger* temperature differences compared to the standard scenario in which they are already present from early times. So these late-time phase transitions would not work.[23]

And we were right: those fluctuations were first observed by the Cosmic Background Explorer (COBE) satellite and announced in 1992, after we had finished most of the work for the paper, but before its publication.[24] (A less self-aggrandizing view of our paper might perhaps be that it was unnecessary, rather than right.) COBE had seen evidence that, only 400,000 years after the beginning of the Big Bang expansion, when the Universe was only about 0.003 percent of its present age, the seeds of all the galaxies in the Universe had already been laid down. The COBE team improved the precision and accuracy of its results over the early 1990s, beginning an age of data-driven cosmology, pairing precise new data with statistical tools developed to take full advantage of them — tools that took full advantage of Bayes' theorem and the accompanying view of probability.

Figure 20 shows the final map of the sky from COBE.[25] It is a view of the sphere of the sky flattened out to show on a page or screen. Unlike many familiar map projections of Earth, which distort the sizes of countries more and more as you move from the equator to the poles, this map (known as a Mollweide projection) gets the area of any region correct, but distorts shapes away from the equator. Figure 21 shows a map of Earth in the same projection (you can see how Greenland, for example, does not look as gigantic as in more familiar maps).[26]

The COBE satellite was followed by a series of CMB telescopes on the ground and dangling from balloons more than 120,000 feet off the ground,

COBE

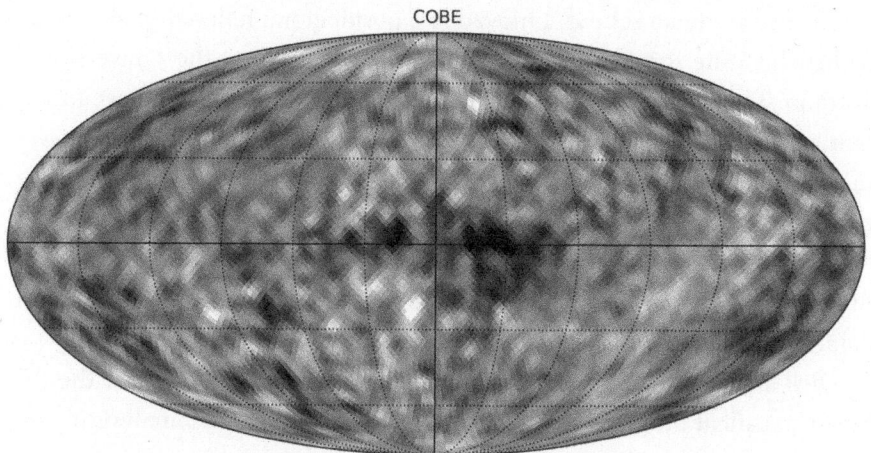

Figure 20. Map of the CMB sky as measured by the COBE satellite.

Earth

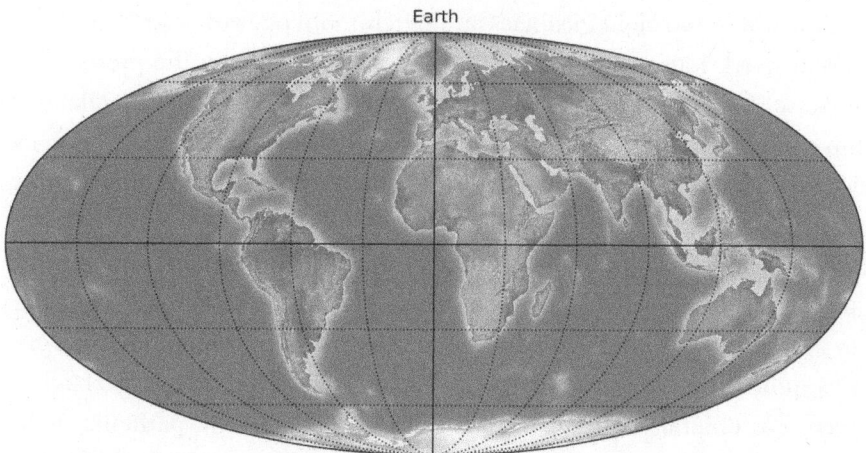

Figure 21. Map of the Earth in the Mollweide projection.

heights that get you some of the advantages of satellites at small fractions of the cost (and with the added advantage that you can sometimes recover—and then repair and improve—your instrument after it flies). The experimental groups developing, building, and deploying these telescopes were keen to share their data with their more theoretical colleagues who had been thinking about how best to analyze their data.

So after graduate school, I moved to a postdoctoral fellowship at the Canadian Institute for Theoretical Astrophysics (CITA), at the University of Toronto. There, I was fortunate enough to work with two of the most important cosmologists of the era—and two of the smartest people I have ever met—Nick Kaiser and Dick Bond. They had spent their careers considering exactly that intersection of probability, theoretical physics, and observational results coming to the forefront of cosmology. I ended up working with them both, separately, on related aspects of the analysis of cosmological data.

Dick and Nick had each made some of the first predictions for the detailed statistical properties of the CMB and in the Bayesian analysis of cosmological data, which were just becoming possible with the computers that were available at the time. (I remember when CITA purchased a two-gigabyte disk, thinking we would never be able to fill up anything so large. Today, my watch has eight gigabytes and much more powerful hardware.)

With Nick Kaiser, I looked at the way galaxies move in the present-day Universe. On the surface, the Hubble-Lemaître law gives us the relationship of the redshift of a galaxy—which is a measurement of its speed—to its distance. However, this interpretation isn't quite right: the law only applies to a perfectly smooth Universe. In the real Universe, full of lumpy, dense objects, a galaxy's speed is also affected by the push and pull of other objects due to gravity and other forces. The deviations from the Hubble-Lemaître law are called "peculiar velocities," and the pattern of those velocities contains information about the history and contents of the Universe. Higher speeds correspond to more matter in the Universe and in particular help measure the total amount of matter contributing to the gravitational field— both normal and dark matter. These Bayesian measurements of the peculiar velocities confirmed, albeit tentatively, the large amount of dark matter necessary to explain velocities on even larger scales than the clusters that Fritz Zwicky observed: we were using measurements of the overall velocities and mutual gravitational attraction of clusters themselves, averaging over the galaxies within them.[27] Nick Kaiser died in 2023, far too young at 68, after having left CITA for Hawaii in the late 1990s, and then for Paris in the 2010s.

But peculiar velocities are notoriously difficult to measure, so Dick Bond and another CITA postdoc, Lloyd Knox, and I worked out how to measure the statistical properties of CMB maps and use those properties to infer the parameters that describe the history and the contents of the Universe. In our first papers together, we analyzed publicly available data from COBE and some of the ground- and balloon-based telescopes using our new methods.[28] This was fun, but mostly we were confirming, in new and more accurate ways, the results that these teams had already published.

As with most scientific endeavors, the hard part isn't getting everything to work (especially if you're confirming someone else's results), but thinking about the ways in which it might fail. And that requires a much deeper understanding of the data. So after that, we started working directly with the experimental teams. For my next postdoctoral fellowship, I went to the University of California at Berkeley, where I joined the collaborations for the MAXIMA and BOOMERANG experiments.

These were balloon-borne telescopes using similar technology but different strategies: MAXIMA (Millimeter Anisotropy eXperiment IMaging Array), which was centered at the University of California at Berkeley, launched from the US National Scientific Ballooning Facility in Texas in a series of roughly eight-hour flights that allowed it to see a patch of about 60 square degrees. BOOMERANG (Balloon Observations of Millimetric Extragalactic Radiation and Geophysics—"geophysics" appears mostly for the appealing acronym!), with principal investigators from Caltech and the University of Rome, flew from Antarctica, which allowed it to take advantage both of the very dry and clear air and the Antarctic polar vortex, an atmospheric phenomenon in which the air circulates in very stable winds around the poles for months at a time. Combined with long-duration balloon technology, BOOMERANG was able to stay in the air for a couple of weeks and thereby map much larger areas of the sky. The experiments had been started by the same core group of scientists, but had diverged over time, using similar hardware but very different strategies. Many of that core group were—officially—members of both teams, but had cast their lot with one or the other of the experiments. Some of us, however, were actively working on the analysis of both sets of data.

A very exciting but stressful time: despite the many scientists in common, the teams were still in competition. COBE had observed the CMB on the largest scales of the sky, which meant it could only give broad confirmation of the cosmological model that produced the fluctuations it observed. By measuring the fluctuations down to scales of a bit less than a degree on the sky (roughly the size of the moon), these experiments could see evidence of sound waves in the primordial plasma on scales larger than galaxies. Photons, protons, and electrons were moving around in response to the gravity of the dark matter, leaving their imprint in the CMB—and giving us an opportunity to measure the properties of that plasma in detail: How much dark matter? How many atoms? How many photons?

In the end, both experiments worked, spectacularly, and produced the first two high-quality small-scale maps of the CMB. Predictions for the statistical properties of the patterns in the CMB had been made, first by Jim Peebles at Princeton, and then solidified in more detail by my CITA mentor Dick Bond with his early collaborator George Efstathiou.[29] These calculations showed that the typical size of spots in the microwave background was around one degree, and experiments down to these scales could access much more cosmological information than COBE.

With the high-resolution results from MAXIMA and BOOMERANG, we were able to confirm these theoretical predictions and make precise measurements of the cosmological model, using the Bayesian techniques we had been developing over the previous decade. Dark matter, which had previously been only indirectly observed, was confirmed on much larger scales than in the galaxies and clusters observed by Zwicky and Rubin.

At these scales, CMB experiments could finally apply the geometrical tests, comparing physical and angular sizes to the most distant objects imaginable—spots in the CMB. This allowed us to get a handle on the geometry of the Universe: the main BOOMERANG paper was entitled "A Flat Universe from High-Resolution Maps of the Cosmic Microwave Background Radiation," although in fact measuring the geometry independently is tricky since we don't have an independent measurement of the *physical* size of those spots at the time at which we see them, 400,000 years after the start of the Big Bang.[30]

In the meantime, other groups were preparing still more ambitious experiments. After COBE, both NASA and the European Space Agency (ESA) started looking for the next generation of CMB satellites. In 2001, NASA launched the Microwave Anisotropy Probe, soon rechristened WMAP, the "W" for David Wilkinson, one of the original quartet of Princeton physicists who had been scooped by, and interpreted the work of, Penzias and Wilson, and had died in 2002. WMAP used similar technology to COBE, but was a decade more advanced and had the advantage of knowing more precisely what to look for.[31]

WMAP confirmed what COBE had found and our MAXIMA and BOOMERANG results—a map of the whole sphere of the sky, measuring tiny temperature fluctuations of angular size less than half a degree; it was able to measure temperature differences down to one-millionth of a degree. The map, shown in Figure 22, has the same overall pattern as the COBE map from a decade earlier, but with much higher resolution and lower noise.[32]

Analyzing this map (using more advanced techniques related to the ones we had developed) allowed the WMAP team to measure even more of the parameters that describe the contents, history, and evolution of the Universe: photons, baryons, and dark matter, expanding for almost 14 billion years.

WMAP's results allowed the team to measure the numerical values of these quantities with precisions of better than 10 percent. This was remarkable for a field that had, until recently, been derisively described as "twiddle astrophysics" for the mathematical "~" sign, usually used to describe a quantity whose error bars are so large that we might only be confident of a value to within a factor of a few: we were confident that the Universe was not 5 billion years old, nor was it 20 billion years old, but we would not have been too surprised if it was 7 billion or 15 billion years old.

The most famous example of this uncertainty was the Hubble constant, the H_0 that appears in the Hubble-Lemaître equation. From the 1960s through the 1990s, some cosmologists measured this number to be about 50 km/s/Mpc (kilometers per second per megaparsec, the strange units that we use to describe the expansion rate of the Universe); others measured the number to be 100 km/s/Mpc, differing, sometimes rancorously, by a full

WMAP

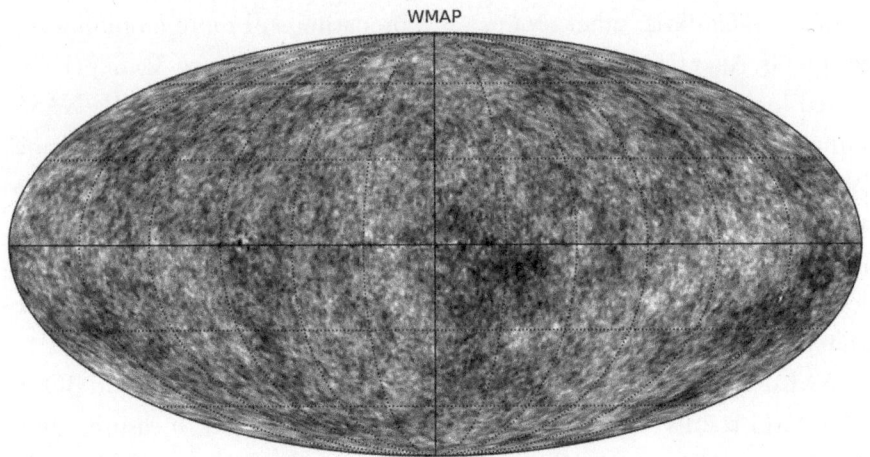

Figure 22. Map of the CMB sky as measured by the WMAP satellite.

factor of two. By the late 1990s, better understanding and measurements of Cepheid stars and other standard candles by the Hubble Space Telescope had narrowed this down to 72±8 km/s/Mpc: neither group was correct.[33] This number was confirmed by the very different methods of WMAP, giving 70±2 km/s/Mpc.

Planck

But even as the WMAP team published its papers, we knew we could do even better. As WMAP was being designed and built, ESA was building its own satellite, so its project was a few years behind NASA's. Originally called COBRAS-SAMBA, ESA's satellite had this slightly unwieldy name that reflected its history as two separate telescope ideas combined into a single satellite with a larger set of detectors. It was renamed the Planck satellite, after Max Planck, the pioneer of quantum mechanics and thermodynamics, who had given his name to the shape that would eventually describe the spectrum of the CMB.[34]

The Planck satellite was more sensitive than WMAP, could detect smaller shapes on the sky, and could see a broader range of colors on the sky ("colors," here, means frequencies of light, but with longer wavelengths than the visible light we normally see). All these improvements were crucial, as the pattern of temperature variations on the sky persist but get fainter as you look at ever finer structures. Perhaps more importantly, Planck's broad range of color sensitivity means that it could more easily distinguish between actual CMB photons and those from so-called astrophysical foregrounds: stars, gas, and dust in the Milky Way and more distant galaxies, contaminating our view of the primordial CMB photons.

The Planck satellite was launched on March 9, 2009, from the ESA's spaceport in Kourou, French Guyana, which is the most equatorial part of the European Union, despite Kourou being in South America (owing to the ongoing and still complicated postcolonial history of Europeans inhabiting other continents). I was fortunate enough to be there. The launch of a rocket is one of humanity's most spectacular technological achievements and one of the scariest—especially when a satellite that you've been working on for nearly a decade is strapped to the top of the giant tank of flammable liquid.

But the launch was a success, and Planck made its way out to orbit a million miles from Earth, in a location from which it could put Earth, the moon, and the sun at its back, and stare out into the cold darkness without contamination from those hot, glowing spheres. To achieve its sensitivity, parts of the Planck detectors had to be cooled down to less than 1/10 of a degree Kelvin—almost certainly the coldest place in the solar system (and possibly beyond) outside of terrestrial laboratories.

And those detectors worked spectacularly. Planck took data from 2009 through 2013 and was analyzed by a team of scientists from all over the world that was led by principal investigators Jean-Loup Puget and Reno Mandolesi, reflecting the genesis of the project as the (mostly) French COBRAS and (mostly) Italian SAMBA proposals.

Data analysis is hard, perhaps less difficult than building and launching a satellite, but certainly less professionalized: more chaotic, dependent upon all the skills of the scientists doing the work. And not just our technical skills are required—everyone involved is well-trained and very smart in their scientific disciplines. Working in a collaboration with hundreds of your fellow scientists requires soft skills for which many or most scientists have never been trained. Astrophysicists, we learned, are terrible at managing and at being managed.

And I certainly include myself in that statement—I was a middle manager in Planck, responsible for measuring aspects of how the Planck telescope distorts our view of the sky, smearing out any structure below a few arcminutes in size. This meant that I had to manage my small team at the London Planck Analysis Centre located at Imperial College, but report back to our scientific managers at the Institut d'Astrophysique in Paris as well as our UK government paymasters, fulfill my local teaching and admin duties, supervise graduate students and postdocs, and try to keep publishing scientific papers while waiting for the completion of the Planck analysis. I also had other Planck duties, such as helping write various papers describing our results, and, in fact, leading some small amount of research with our data. I never really got the hang of juggling all these responsibilities and deadlines, all with their own lines of control, and I am sure that I was not alone.

But the effort worked. More than a decade after its launch, we published the final tranche of Planck papers in 2020, our last meetings held remotely during the COVID pandemic lockdown. We measured those cosmological parameters that describe the history and evolution of the Universe on the largest scales. Before COBE, some cosmological parameters were not known to better than a factor of two. WMAP's results gave us the parameters to 10 percent. And Planck measured them roughly a factor of ten better.

Figure 23 shows the map of the sky that enabled these fantastic results, a map with much higher resolution and lower noise than that of WMAP—but hard to reproduce on the small scales I can show here.[35]

It's difficult to be precise about the improvement that Planck offered over WMAP. As experiments get more precise, we expand the space of models that we are considering, models not only of the cosmology that we care most about, but models of how our data are affected by noise from our experiment and contaminated by other sources of light that come into our telescope, the glow of gas and dust from our solar system, the Milky Way and distant galaxies, and cosmic rays that glitch our detectors. Our better under-

Figure 23. Map of the CMB sky as measured by the Planck satellite.

standing of these effects lets us (forces us to) be more realistic when we interpret our data and when we combine it with other measurements. One measure of the improvement is just to count the number of pixels in the maps produced by the experiments: COBE's map had about 6,000 pixels, WMAP 3 million, and Planck 50 million.

Almost 30 years after COBE first announced its results, the era of precision cosmology, as well as the era of cosmology as big science, was fully here. But we need accuracy along with precision, and high precision makes us more sensitive to problems with our data, or with the model.

ELEVEN

The Inflationary Universe

I have been describing the different epochs of the Universe's evolution by their times: 3 minutes for the formation of the light elements in Big Bang nucleosynthesis; 400,000 years for the formation of the CMB; a current age of 14 billion years. But how am I defining the start—time zero? When and how did the Universe begin?

Sakharov's insights, and the detailed work of a century of scientists studying general relativity and cosmology, have given us a model in which the Universe grows more complicated, and more interesting, over time. The very early Universe was dense, hot, and smooth. Matter then differentiated itself from antimatter, protons and neutrons formed from quarks, nuclei formed from those protons and neutrons, atoms formed from the primordial plasma, and, eventually, objects condensed out when gravity was finally able to take over.

But this timeline, coupled with the precise measurements from Planck and other twenty-first-century experiments, implies a very specific beginning for the Universe. The early Universe must have been very nearly uniform over vast distances: from one side of the sky to the other, the CMB temperature is constant to about one part in 100,000. That is, the entire observable Universe seems to have been in thermal equilibrium when the CMB was formed, at about 400,000 years. But this is a puzzle: it takes time to come to thermal equilibrium. If I put a tray of water in the freezer, it doesn't freeze immediately. Temperature is a description of the random motions of particles, and the speed limit on the transmission of those random motions is given by the speed of sound in whatever stuff we're describing. In our case, that "stuff" is the plasma of the early Universe, made up mostly of electrons, protons, and many, many photons. (Recall that there are about a billion photons for every electron.) When there are so many photons around—so many particles of light—the speed of sound is a bit more than half of the speed of light. Fast, but not infinite.

So, if the Universe is about 400,000 years old, we might expect patches roughly 200,000 light-years across to have come to the same temperature (that is, the distance that a sound wave could have traveled, moving at about half the speed of light). But that distance, at the time of the formation of the CMB, corresponds to only about one degree on the sky.

But there are about 40,000 patches about one degree across over the whole sky; there should not have been time for them to come to equilibrium and certainly not to have done so to better than one part in 100,000.

The Universe evolves in other ways, too. As it expands, the geometry changes, moving further and further from being flat over time. Today, we know that the Universe is very, very close to flat: cosmological experiments such as Planck indicate that Ω, the number that describes the geometry, is within about 1 percent of $\Omega=1$. At the time of the formation of the CMB, for example, it would have been about 400 times closer to flat than it is today, and about 10 million times closer to flat at the time of BBN, three minutes after the Big Bang.

These numbers are hard to understand: Why was the Universe so finely tuned to start out perfectly flat and in perfect equilibrium? It is certainly possible to answer that the Universe was just made that way—these initial conditions are simply a part of the description of the laws of physics, perhaps not in need of any further explanation. Or perhaps they have to do with a more fundamental theory that we don't yet understand. This is a perfectly respectable response, but one that rankles many physicists who are eager to explain the way things are, not just describe them.

A Bigger Bang

So, perhaps, we can explain this early state of the Universe with the laws of physics that we do know. In the 1970s, physicists were starting to think about the behavior of the expanding Universe if it were filled with exotic matter.

They had begun to think about the Higgs field, a version of the famous Higgs particle postulated in the 1960s by Peter Higgs, which quickly became crucial to the standard model of particle physics but was only observed at CERN in 2012.

In the hot and dense early Universe, the Higgs, or other fields like it, might behave in very different ways, not like individual particles, but rather as a medium pervading the whole Universe. If that medium is constant over some region, its density dominating over all other kinds of particles, such so-called scalar fields exert a sort of negative, repulsive gravity, driving the expansion of the Universe to accelerate over time. This is very different from a Universe dominated by any other kind of matter, be it photons, normal matter particles like protons or electrons, or even dark matter; the attractive gravity of any of these particles cause the expansion to slow down over time.

We can define the *horizon* in an expanding Universe as the distance that a particle could have traveled since the start of the expansion. Roughly, this gives the size of the observable Universe at any time, which is also the maximum distance that a signal could have traveled over that time. We say that a patch of that size is in *causal contact*: particles within the horizon could have had time to interact. In a universe whose expansion is slowing down, the size of the horizon is always growing: the longer you wait, the more you can see.

But the ever-accelerating expansion caused by these scalar fields causes the horizon to shrink: the distant regions of the Universe are expanding so rapidly away from you that you lose sight of them. We call this possible early period of accelerated expansion *inflation*.

This concept is exactly what is necessary to solve the puzzle of the initial conditions, if inflation goes on for long enough. A shrinking horizon is analogous to zooming into a curved surface—if you look at a small enough patch, it will look smooth and flat. This solves the flatness problem, driving the value of Ω very close to the value corresponding to a flat Universe, $\Omega = 1$, close enough to that value that we would still expect the geometry to be indistinguishable from flat even today.

By shrinking the horizon over time, parts of the Universe that had already been in causal contact fall out of contact with each other. If those parts had already been in thermal equilibrium at a particular temperature, they will stay at the same temperature even if they are out of contact.

For inflation to work, it must last for a long time, at least 10^{-33} seconds. That's just one decillionth of a second, but at this very early time in the history of the Universe, it is long enough for the size of the Universe to increase by a factor of more than 10^{27} (1 followed by 27 zeros—an octillion). These are just the minimum amounts of expansion required to make the Universe sufficiently flat—an even longer period of inflationary expansion would have the same effect.

The outline of this idea, that accelerated expansion could ameliorate these problems with the initial conditions of the Big Bang, was discovered more or less independently in 1979–1980 by Alexei Starobinsky in Russia and Alan Guth at MIT in the US. The early versions had some issues with exactly how inflation stopped happening: we know that the success of the *hot* Big Bang theory implies that the Universe has had a very long period of decelerating expansion, responsible for all the transitions that led to the formation of nuclei, atoms, and eventually the formation of galaxies and clusters of galaxies.

As the Universe inflates, however, it cools down—this is exactly how refrigerators work, rapidly expanding a gas to lower its temperature. To restart the hot Big Bang, the Universe needed to reheat, converting the scalar field responsible for inflation into the kinds of particles that would evolve into today's Universe.

Guth's original inflation had a particular model for reheating, as different parts of the Universe with different values of the scalar field crashed into each other. But the community quickly realized that this would not work: these distinct areas would be too far apart to collide. The model that fixed this problem is now called *slow-roll* inflation, developed first by Andrei Linde in Russia and Andy Albrecht and Paul Steinhardt in the US. Inflation does not end by crashing but when the scalar field gradually settles down to a smaller value and no longer dominates the Universe, returning it to its decelerating expansion.

In this slow-roll model, the field responsible for inflation has the chance to decay into the particles of the standard model that we see today. The Universe reheats because these particles are created as high-velocity pairs, which interact with their neighbors and quickly find thermal equilibrium, giving us the hot Big Bang that we need.

Inflation takes an early universe that may not be flat, may not be everywhere uniform, and may not be hot, and converts it into a hot Big Bang, in thermal equilibrium, with a single, constant temperature. If inflation only gave us these properties, that would have been enough to make it a viable theory for understanding how the Universe began.

But soon after Starobinsky and Guth and the relatively quick birth of slow-roll inflation, cosmologists realized that inflation had another important effect, when the details of quantum mechanics were properly taken into account. Quantum mechanics predicts randomness at the very smallest scales. Inflation takes small scales and stretches them out. Putting these two facts together implies that those small-scale quantum fluctuations will inflate to what are today large enough to be observable by astronomers, when they become, over the expansion history of the Universe, actual differences in the density of matter from place to place.[1]

The physics of inflation doesn't just tell us that these density fluctuations are random, it gives a recipe for calculating their distribution: the probability that you will find a fluctuation of a given size, at some particular distance from a fluctuation of some other size. That information can be distilled into a single curve called a *power spectrum*.

This power spectrum, in turn, is exactly what is needed to describe those fluctuations with a more complicated version of the Gaussian distribution that we encountered earlier (since we need it to describe the statistical properties of not just one quantity, but the density of matter in the entire Universe). Inflation gives the power spectrum left over after the time of reheating, which we call the *primordial* power spectrum; our cosmological model for the ensuing hot Big Bang tells us how that power spectrum will evolve over time. This lets us predict the statistical properties of the cosmic microwave background, as well as of galaxies in the Universe today.

Even in the early 1980s, prior to the discovery of fluctuations by COBE, it was known that the fluctuations in the CMB would be very small. However, the simplest models put forward in this first set of papers typically predicted fluctuations so large that they would already have been observed with the technology available then. But they did predict that the *shape* of the power spectrum function was more or less as observed in the distribution of galaxies on the sky at that point.

This led to the conclusion that the Higgs field itself could not be the scalar field responsible for inflation, but instead required dialing in a new kind of particle, typically called the "inflaton" and denoted by the Greek φ (phi).

This new particle allowed a lot more freedom in modeling inflation, including its mass and its interaction with itself and with other particles. It became clear very quickly that the inflaton could have a very wide variety of such properties and still result in an inflating Universe with fluctuations in accord with observations.

This, then, is "The Random Universe," a description of our theory of cosmology as well as the title of this book. The large-scale properties of the Universe—its curvature, density, and temperature—are determined by the accelerating expansion of inflation. Superposed on those large-scale properties are tiny quantum mechanical fluctuations, stretched out from subatomic scales eventually across the visible Universe, with a specific prescription determining their statistical properties.

This is what COBE's results hinted at. Figure 24 shows possible power spectra of the CMB for models that all would have been allowed by COBE's observations.[2] The curves show the different angular scales on which the CMB sky shows structure, vibrations in the primordial plasma of baryons and photons as the Universe evolves.

These were solidified by BOOMERANG and MAXIMA, WMAP and Planck, which winnowed the allowed models down into today's standard model of cosmology. The error bars on one of the curves show the Planck data, rejecting most of these models that were viable after the COBE measurements in the 1990s. The tiny error bars, barely visible compared to the width of the points, only allow a small subset of possible cosmological parameters to be compatible with the data.

These more recent experiments simultaneously measured the shape of the primordial power spectrum and the physical processes responsible for evolving those fluctuations from the end of inflation until their observation today.

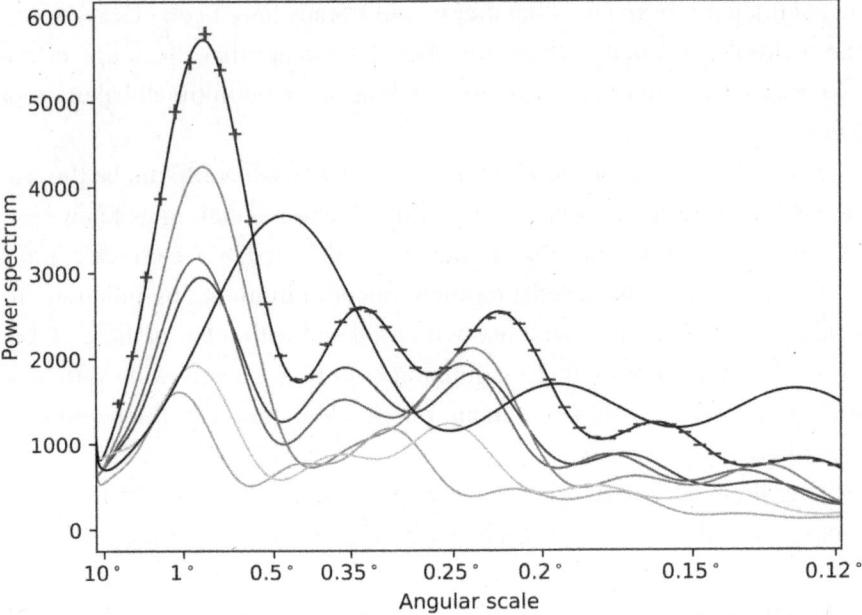

Figure 24. The CMB power spectrum, as measured by the Planck satellite, along with power spectra for many other models that had been deemed possible given earlier data but that are now ruled out. The nonuniform horizontal axis shows the typical angular scale at which the power is measured.

Gravitational Waves?

But inflation makes further predictions beyond those that I've discussed. The same quantum mechanical fluctuations that eventually grew into the large-scale galaxies and clusters that we observe today are also the source of *gravitational waves*: these are fluctuations in space and time, traveling at the speed of light. Gravitational radiation is a prediction of Einstein's general theory of relativity and was conclusively observed by the LIGO experiment in 2016.

LIGO observes gravitational radiation from violent motions of black holes and other so-called compact objects like neutron stars spiraling inward toward each other, all in our Milky Way or external galaxies. On Earth, the signals from these events are typically less than one part in one quadrillion—so LIGO needs to measure changes in distance comparable to much less than the size of the nucleus of an atom. This is a spectacular technological trick, literally performed with mirrors and laser beams.[3]

The gravitational radiation from inflation is somewhat different: it doesn't come from individual sources, but from wiggles in space-time interacting with the inflaton field. These fluctuations in gravitational radiation are described by a primordial power spectrum stretching across the observable Universe, analogous to the primordial density power spectrum that describes the initial conditions for the pattern of galaxies over the Universe. But the pattern of gravitational radiation is most easily visible not in the overall brightness of the CMB, but rather in another property called *polarization*.

Polarization accounts for the fact that light is described not only by its overall brightness, but also by a direction. For many sources of light, the individual photons are randomly oriented, but some processes produce polarization more in one direction than others. This occurs, for example, when light reflects off a mirror or water. So when you look at a lake on a sunny day, if you are wearing polarized sunglasses—made so that polarized light passes through in one direction—they cut down the glare of the polarized light from the surface and let you see the *unpolarized* light reflecting off objects beneath.

So now we build our CMB detectors with, essentially, a set of polarized sunglasses sitting in front of each of them, allowing us to be sensitive to only certain directions of the polarization of the light. And one of the sources of the polarization of the CMB is gravitational radiation, causing motion in the primordial plasma that is translated into polarized light. Moreover, this polarization has a very specific pattern of swirls that we can separate out.

If we can convincingly detect these swirls, and convincingly distinguish swirls due to the early Universe from other sources, then we can detect this inflationary gravitational radiation. Neither the WMAP nor Planck satellites had managed this feat.

But in 2014, this is exactly what the BICEP team thought they had done. BICEP was a ground-based CMB telescope located at the South Pole with one of the most sensitive polarization detectors ever built. The team carefully arranged their observations to concentrate on the cleanest parts of the sky, areas chosen so that there would be few other sources of polarized light from, for example, gas and dust in our own galaxy.

Because they had so carefully designed their observation strategy, the team thought they would not have to do much work to distinguish the inflation signal from any so-called astrophysical foreground. And they saw the swirly pattern indicative of gravitational radiation from inflation![4]

The results made a splash, with a press conference, articles in major newspapers, television and radio coverage (I spent about two hours doing a "deep background" interview with a BBC science correspondent from Loch Lomond in Scotland, where I had been walking at the time with my wife). The team made a reality TV video of the BICEP investigators surprising Andrei Linde, one of the cocreators of the theory of cosmological inflation, with a bottle of champagne at his California house to celebrate the empirical proof of his theory. It was one of the most exciting days for cosmology in recent memory. (There had been a big press conference the previous year for the release of Planck data, too, but we were reporting confirmation of what had already been observed, rather than a new discovery.)

Except. The team had misinterpreted their data. The signal that they observed was indeed due to dust in our galaxy, rather than inflation. Modeling the galaxy is hard, in some ways harder than modeling the Universe as a whole on large scales. They had assumed that by looking in a direction that has only a small amount of dust, they could ignore its contribution to the polarization of the CMB. However, that doesn't always work: often, less total dust can lead to more polarization, as there aren't as many opportunities for the dust to randomly reorient the polarization of the light that shines through it.

And, indeed, the initial results reported by BICEP were already in some contradiction with Planck's results from the previous year. We had, somewhat tentatively, put an upper limit on the amount of gravitational radiation from inflation that was higher than BICEP's results.[5] This limit was tentative because we did not report a polarization measurement that year; rather, we could measure gravitational radiation by its small contribution to the temperature on the sky. But even this was enough for us to put a limit on gravitational radiation that was twice as high as BICEP's purported detection; the way around this contradiction was to postulate a somewhat unusual form of inflation that gave a different signal in those two different measurements.

By the end of 2014, we on the Planck team had released a paper that gave a more precise measurement of the polarization due to dust, and the following year the Planck and BICEP teams collaborated on a paper, putting the strongest constraint on gravitational radiation to date.[6] This was an incredible disappointment (and perhaps a small embarrassment) for the BICEP team.[7]

But it was also an opportunity to see the interaction of models and data—the scientific method—in action. The BICEP team had a fantastic set of data from their telescope, and they interpreted it in light of a particular model of the Universe, one that allowed for inflation but did not correctly account for contamination by dust. The Planck team had a separate set of data that partially contradicted this; only by combining the two sets of data, and making more sophisticated models, could we understand inflation coherently.

In the decade since 2014, the Planck team finished the analysis of its full dataset, with still no detection of gravitational radiation. But there is a new generation of experiments preparing to measure the polarization with yet more precision. The BICEP team has deployed more, and more sophisticated, telescopes—and a new satellite, developed by the Japanese Space Agency, JAXA, is expected to launch in the early 2030s.

For me, the most exciting prospect is a set of several telescopes called the Simons Observatory, with many tens of thousands of individual detectors (the Planck satellite had about 50), deployed in the Atacama Plain in Chile, one of the highest and driest sites in the world.

Will the Simons Observatory see gravitational waves? Of course, we don't know. It is entirely plausible that inflation did occur, but in some models the predicted background of gravitational waves is so weak that it will be impossible for us to observe it no matter how precise our measurements. But the Simons Observatory is already starting to make its measurements, and we should have an answer over the next few years.

I can't wait.

TWELVE

Tension or Concordance?

The hot Big Bang model has been refined over the last century as scientists have developed new theoretical ideas to incorporate new observations. But we aren't finished yet. The model still has many unexplained parts and even some internal contradictions.

The most obvious unexplained part is the identification of *dark matter*. Dark matter was first hypothesized by Zwicky to account for the motions in clusters of galaxies, and then by Rubin to explain the motions of stars within galaxies themselves. But it is also necessary on the very largest scales: neither the pattern of the CMB nor the pattern of galaxies makes sense if there is no matter other than gas, and dust, and stars.

The problem with dark matter is exactly the "dark" part—it interacts hardly at all with the atoms that make up everything that we can see more directly. But this means that it is very hard to create dark matter in a particle accelerator or somehow observe it in a laboratory. On the other hand, quantum mechanics implies that it's very difficult for a particle to have *only* gravitational interactions with other particles.

Understanding the nature of the dark matter would not only help complete the cosmological model. It would also add a new entry to the bestiary of quantum particles, which would help our colleagues in particle physics complete their models as well.

It's Dark Out There

And there is another "dark" component to the Universe, known by the cosmological term of art *dark energy*. Dark energy was, in fact, a part of Einstein's theory of gravity since the very beginning, although it was only given its "dark" name in the 1990s.

Originally, dark energy was enlisted to avoid a Big Bang model. In the 1910s, the expansion of the Universe was not yet understood, and instead Einstein searched for a cosmological model that would not change over time. To counteract gravity's usual attraction, his model had to be balanced by a separate force pushing things apart. His dark energy, known also as the "cosmological constant," and usually denoted by the Greek Λ (lambda), provided exactly the necessary repulsion.

But once the expansion of the Universe was observed by Hubble in the 1920s, this term was no longer obviously needed in Einstein's equations, and the cosmological constant was, mostly, eschewed. It remained a possible part of the theory, however, and would occasionally be dragged out to solve some small problem in understanding cosmological data.

By the 1990s, however, some of those supposedly small problems were not going away. One egregious problem resulted from trying to combine measurements of the Hubble constant (H_0, the expansion rate of the Universe from the Hubble-Lemaître law) with measurements of the ages of the oldest objects in the Universe (under the obvious assumption that the Universe as a whole must be older than the objects in it!).

In a hot Big Bang Universe as described originally by Friedmann, Lemaître, Robinson, and Walker, the product of the Hubble constant and the age of the Universe depends only on its overall geometry, along with the density of its constituents (the Ω parameters). So measurements of the Hubble constant and the Universe's age can combine to give an indirect measurement of the geometry and density of the Universe.

In 1995, astronomers were getting really good at determining the ages of stars, especially the ages of clusters of stars all born at the same time, which allowed us to combine the noisy measurements of the ages of many thousands of stars. This technique was applied by one of my colleagues at the Canadian Institute for Theoretical Astrophysics, where we were both postdocs at the time. He found an age of the Universe of almost 15 billion years.[1] At the same time, measurements of the Hubble constant were beginning to settle down at a value of somewhere between 65 and 80 km/s/Mpc. Combining those measurements, even with their error bars, and with the restriction that we don't know how long after the start of the Big Bang the stellar clusters were born, allows us to put some constraints on the geometry and constituents of the Universe.

Using Bayes' theorem, we could in principle use this information to determine the amount of matter and dark energy in the Universe. Unfortunately, because the age measurement is really just a limit on the age of the Universe as a whole (it must be more than about 15 billion years old), these results do not translate to particularly useful constraints. Instead, we can use the information not to measure those parameters, but to compare different models for the Universe. In a model with dark energy (in the form of Einstein's Λ or otherwise) causing accelerated expansion at late times, the Universe can be older while the value of H_0 stays the same.

Even these relatively meager data—large error bars on the Hubble constant and a lower limit to the age of the Universe—were enough to show us via Bayes' theorem that a model with at least some significant amount of dark energy was strongly favored over the alternatives.[2] Here, I used Bayes' theorem not only to measure parameters, but to compare entire models.

One such model, the most popular among theoretical cosmologists at the time, has normal and dark matter, zero dark energy, and a flat geometry. This model was often called "standard cold dark matter" or "CDM" for short. This model is particularly simple: it has no free parameters. A second model keeps the Universe flat, but has comparable amounts of matter and dark energy, and so has one free parameter, either the total amount of dark energy or of (normal plus dark) matter. This is often called "ΛCDM" because of the mixture of matter and dark energy.

Using the full machinery of Bayes' theorem, I found that the most probable model turns out to be ΛCDM: a flat Universe consistent with inflation, but with dark energy like Einstein's cosmological constant. Around the same time, there were also tentative, if somewhat indirect, measurements of the densities themselves, indicating that there was not enough normal and cold dark matter in the Universe to account for a flat geometry. However, the results from COBE gave some evidence that the geometry of the Universe was at least close to flat and consistent with inflation. Simultaneously, astronomers were beginning to see evidence that the expansion of the Universe has been accelerating for the last several billion years. Because the Universe has been expanding faster, the same objects are dimmer than they would be absent acceleration. By the late 1990s, teams of cosmologists had realized that a certain kind of exploding star, type Ia supernova (or "SNIa"), would make an excellent standard candle, observable out to very large distances.

The extra dimming of these objects was beginning to be observed in the 1990s (and these observations would eventually result in the award of the 2011 Nobel Prize), and could be quantitatively converted into a combination of the amounts of matter and dark energy. This model was first dubbed "concordance cosmology" for its ability to simultaneously account for a host of evidence, from ages to CMB fluctuations to the brightness of standard candles.[3]

To show this evidence, we often plot the posterior probability of the parameters for a model, given our observed data, in two dimensions as a *contour plot*, in which the probability is constant along each of the plotted curves, like a contour map with lines of constant height. The innermost curves show the region with the highest probability, enclosing more and more as we move further away from the peak. Figure 25 shows the state of the art in 2001: the combined probabilities from BOOMERANG and MAXIMA, along with results from the first SNIa observations. The two sets of curves overlap at this area of concordance, the ΛCDM model: roughly 30 percent matter, 70 percent dark energy.[4]

Figure 26 shows a more modern version of the same plot, using data from the Planck satellite. Now, the CMB data alone are enough to focus down to the ΛCDM model, with even more precise measurements of the parameters.[5]

In this ΛCDM scenario, the cosmological constant has started to affect the Universe only in the last few billion years or so (compared to its age of about 14 billion years), causing its expansion to accelerate ever faster. Right now, the cosmological constant contributes about twice as much as the

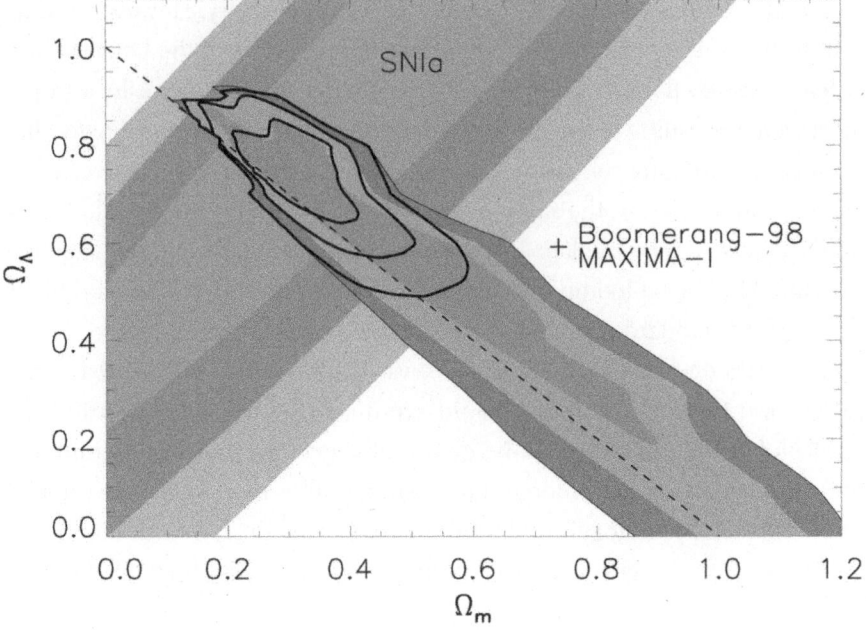

Figure 25. The posterior probability for matter and dark energy from the combination of the MAXIMA-1 and BOOMERANG-98 data, as well as from other data measuring the expansion of the Universe using a supernova (labeled "SNIa"). The diagonal dashed line shows the flat Universe expected from inflation and favored by the combined measurements. (A. Jaffe et al., "Cosmology from MAXIMA-1, BOOMERANG, and COBE DMR Cosmic Microwave Background Observations," *Physical Review Letters* 86 [2001]: 3475–3479.)

combination of normal atoms and dark matter, but as the expansion progresses, the cosmological constant will dominate ever more into the future.

This repulsive gravitational force might sound familiar from the discussion of inflation, which has exactly the same effect. This gives another way of thinking about the cosmological constant: it might be a new field like the inflaton, but with properties that cause it to dominate the Universe not in its very earliest moments, but much more recently. Sometimes this new field is called "quintessence," not from its modern meaning of "perfect example" but from the original Latin meaning of "fifth essence"—a new substance beyond the supposed elements of earth, air, fire, and water.[6] Since then, the term "dark energy" has caught on as a catch-all to describe whatever is responsible for the latter-day accelerating expansion of the Universe.

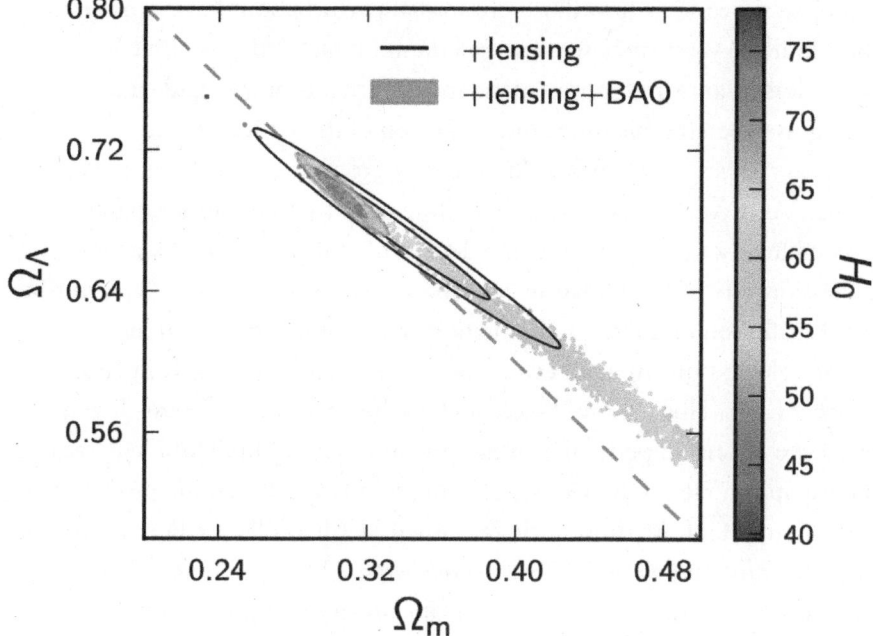

Figure 26. Posterior probability for matter and dark energy from Planck, almost two decades after MAXIMA and BOOMERANG. The different ellipses show Planck in combination with different data available at the time of publication of these results. The gray points show how the Hubble constant changes as the matter density increases. The diagonal dashed line shows the flat Universe expected from inflation, the outcome still favored by the more accurate data. (Planck Collaboration, A&A 594 [2016]: A13, reproduced with permission © ESO.)

Do we have a physical model for dark energy, or is it just a universal constant, like the speed of light, or Newton's gravitational constant? Dark energy in a form similar to Einstein's cosmological constant may arise in the solution to one of the most challenging problems in theoretical physics, the link between quantum mechanics and gravity.

A Golden Age?

In the quarter century since the birth of the ΛCDM model, data have continued to pour in, with the WMAP and Planck satellites refining our picture of the CMB, and measurements of the large-scale pattern of galaxies in the

Universe showing us how those primordial perturbations have evolved over the 14 billion years since we observed them in the CMB; we also have a better understanding and more precise measurements of standard candles, enabling us to see the history of the expansion of the Universe.

So perhaps we are in a golden age of cosmology: we have an overall model, ΛCDM. We have data that allow us to measure all the parameters needed to describe that model: the density of matter and of dark energy, the expansion rate of the Universe, and the numbers that we need to describe the distribution of the small fluctuations remaining after inflation.

We selected this model because of the "concordance" of its fit to a wide variety of data. However, as those data have become more precise, discrepancies have arisen, in particular in the measurement of the Hubble constant, the expansion rate of the Universe. Using standard candles, the most precise measurement is from the amusingly named SH0ES collaboration, giving the Hubble constant a value of 73 ± 1 km/s/Mpc.[7] This is a more or less direct measurement of the expansion rate via the apparent brightness of, and therefore the distance to, the standard candle. But the measurement is not quite direct: we don't have complete information about our standard candles.

A more appropriate name for these kinds of objects is *"standardizable* candles." There are two problems with using these objects. First, the standard candles are not all the same — they are not really "standard." The brightness of one of Leavitt's Cepheid variable stars depends primarily on how fast it varies. The SH0ES collaboration uses a type of exploding star called a "Type Ia supernova," whose brightness depends on how quickly the explosion brightens and dims. In both cases, the brightness depends also on the details of the composition and evolution of the star and its host galaxy. All these relationships need to be calibrated from the same data being used to measure the Hubble constant.

Moreover, in neither case do we have a useful prediction for the brightness of the Cepheid or the supernova. Instead, we calibrate the Cepheids from objects for which we have yet more direct measurement of the distance, and in turn use the Cepheids to calibrate the supernovae. This sequence of measurements, bootstrapping from local to cosmological measurements, is known as "the cosmic distance ladder."[8]

Once we standardize our candles, the relationship between their distances from Earth and their redshifts tells us the value of the Hubble constant. If we trust the ladder of standard candles, we trust the result.

We can also determine the Hubble constant with measurements of the CMB, but in a less direct fashion. The distribution of fluctuations in the CMB, as characterized by the power spectrum (see Figure 24), is separately sensitive to both the initial fluctuations laid down by inflation and the other parameters that describe the Universe, such as the curvature of the Universe and the densities of dark matter, of atoms, and of dark energy, each of which changes the shape of the CMB power spectrum in a way that we can calculate. We can combine those individual parameters in a way that allows us to measure the Hubble constant.

But the indirect nature of this measurement implies that we cannot point to any particular feature of the power spectrum, such as the height of one of its peaks, for example, and directly determine the Hubble constant (or most other parameters that we might wish to measure). Instead, we must calculate the power spectrum over all possible sets of the parameters and figure out which are the best fit to our data, using the mathematics of Bayes' theorem.

We see in Figure 27 another contour plot showing how the measured value of the Hubble constant is related to our measurement of the amount of dark matter in the Universe, considering data from various experiments. Most of the plots agree with one another—the shapes overlap. The most probable value of the Hubble constant from the Planck satellite—represented in the figure by the smallest, thinnest set of curves—is about 67 km/s/Mpc, with error bars about ±0.5 km/s/Mpc. So it might be a bit higher; if so, from the shape of the shaded region, tilting from top left to bottom right, it also becomes more probable that the amount of matter must be lower.[9]

This indirect way of measuring parameters such as the Hubble constant allows an independent check of the assumptions of our ΛCDM model: because of the complicated relationship between the parameters and the shape of the power spectrum, not all possible curves are allowed. If we were very wrong about the model, we would not expect to find the ΛCDM shape, with all its bumps and wiggles in just the right places. So the very fact that we can find a curve that goes through the error bars of our measurements is

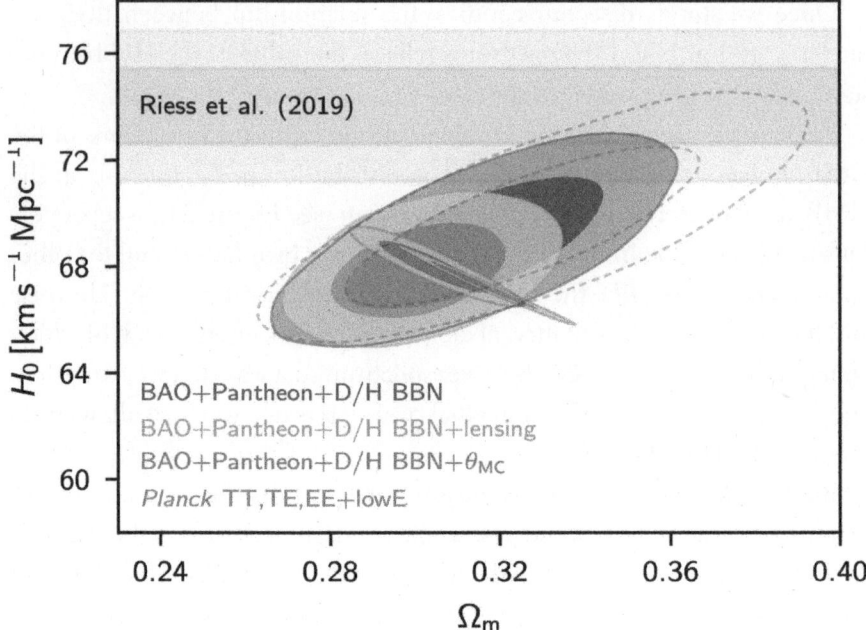

Figure 27. Posterior probability of the Hubble constant and amount of matter in the Universe. The different ellipses show different data available at the time of publication of these results, with the smallest giving the results from the Planck satellite and the horizontal stripes showing the somewhat discordant results from the analysis of supernova data using results from A. Riess et al., *Astrophysical Journal* 876 (2019): 85. (Figure credit: Planck Collaboration, *A&A* 641 [2020]: A6, reproduced with permission © ESO.)

itself a mark in favor of our model. It might not be exactly right, but it must capture a lot of what is going on (recall George Box's aphorism: "All models are wrong, but some are useful").

We saw earlier that measurements of the Hubble constant from the Hubble Space Telescope and WMAP were in good agreement in the first decade of the 2000s. But you might have noticed that I mentioned the value of the Hubble constant according to Planck is 67±0.5 km/s/Mpc, whereas the standard candle measurement was 73±1 km/s/Mpc, shown as a horizontal band on the plot. These are now very far apart compared to the size of the quoted error bars, with a probability of less than one in a million that we would see independent measurements of the same quantity be at least this far apart.

This discrepancy, with its associated low probability, has been dubbed the "Hubble tension": two high-quality measurements of H_0 give different answers. Something in our model must be wrong. But what?

The most obvious place to look is the experiments themselves: observational cosmology is *hard*, and there are lots of ways to go wrong. We could be wrong in our understanding of the standard candles that are used to determine the distance to the galaxies expanding away from us. In practice, we do not have a detailed theoretical model for those candles; instead we must calibrate the model using the very same data with which we measure H_0. This makes it difficult to rigorously and unambiguously check all the assumptions of the model. Imagine, for example, that the supposed standard candles get brighter, slowly, over the history of the Universe. More distant objects would then be dimmer, but in a way that could be very difficult to distinguish from a different value of the Hubble constant. Or perhaps we have misunderstood the contamination of Planck's measurements by gas and dust in the Milky Way, obscuring our view of the CMB? Or we don't understand the hardware of our satellites and telescopes as well as we have thought.

These scenarios require some amount of bad luck (or a conspiracy): such so-called systematic errors would not be expected to give curves that manage to look exactly like those from the assumed ΛCDM cosmology. They might, for example, add extra bumps and wiggles to the CMB power spectrum, or kinks in the Hubble curve. The lack of such extra structure gives strong constraints on exactly the form of any possible systematic error.

This motivates a more intriguing, and exciting, possibility: What if ΛCDM is wrong (or, more precisely, given its successes so far, not quite right)? The CMB measures the Hubble constant by looking at the structure of the Universe almost 14 billion years ago, only 400,000 years after the start of expansion. In comparison, the standard candles rely on the motions of galaxies over the last billion years or so. The expansion rates at those vastly different times are related to one another by the ΛCDM model itself and in particular by the assumption that the Universe is made up of normal matter, dark matter, and a simple form of dark energy like Einstein's cosmological constant. If we relax that assumption—in particular, if we allow a more complicated form of dark energy, dominating the Universe at a time well after inflation, but well before the CMB is released at 400,000 years—we can go some way toward preserving both the standard candle and CMB measurements, at the price of a more complicated model.[10]

Is that a price we are willing to pay? We have complicated the model in a way that has few observational consequences beyond ameliorating this "tension." Whereas dark matter and inflation each solve multiple problems whose solutions were needed to understand a variety of cosmological data, the only prediction of this "early dark energy" is to make sure that the Hubble constant as measured by methods that rely on the distribution of the large-scale fluctuations in the CMB differs, slightly but significantly, from more direct local measurements of the expansion of the Universe. Perhaps Lakatos would consider this an example of a "degenerating research program," if we are willing to accept changes that have no effect other than to reconcile some contradictory evidence.

In fact, the latest information, as of the middle of 2024, is just beginning to favor the possibility that our understanding of some of the local standard candles might be just wrong enough to account for the difference. If we change one rung of our cosmic distance ladder, we might be able to bring our measurements back into agreement. Stay tuned.[11]

However, this discrepancy in measurements of the Hubble constant is not the only evidence that the ΛCDM model is incomplete or subtly wrong. From the very first measurements of CMB fluctuations by COBE, cosmologists have been trying to see if the pattern of the CMB sky is consistent with the predictions of our cosmological model.

In many ways, the particular CMB sky as observed by COBE, WMAP, and Planck satellites does seem unlikely, given our expectations from the ΛCDM cosmological model. If we look at the fluctuations at relatively large scales on the map—structures that stretch most of the way across the sky—the picture seems inconsistent with some predictions of the model. The shapes that we see in the map seem to be somewhat more aligned with each other than we would expect from the model, which has no preferred direction.

But it is very difficult to assess this quantitatively. Our model assigns a probability to any map, no matter what the pattern looks like. There are an infinite number of possible maps, so the probability is, strictly speaking, zero, for any specific map (i.e., where the temperature at every position is exactly specified). Instead, we can ask for the probability of a set of maps that is, in some sense that we need to define, *similar to* a particular map—this will have a finite probability.

In this case, we have to assign a number that describes the lack of isotropy in the map: a quantity that we could calculate for any possible map. We can then ask, if our model is true, how would this number be distributed—what is its probability? Any individual, infinitely precise value will have zero probability, so instead all we can really do is look at where it lies in the distribution, typically approximated by one of our bell-shaped curves. Is the actual observed value near the peak—and hence likely—or out in the low-probability tails? To be somewhat more precise, it is traditional to describe these measurements by the total probability of all the ways that our number could be as unlikely or *more unlikely* than the observed value. This is the well-known, or infamous, p-value from traditional statistics, but it is difficult to interpret as a probability in the realm of Bayesian model-building.

These low-probability patterns are collectively referred to as "anomalies," and a small industry of cosmologists has sprung up to further quantify, and, possibly, explain them.[12] My favorite possibility is that the simplest model of the Universe that allows just one of three possibilities—flat, spherical, or saddle-shaped—is not quite right. Those are the three candidates for the *curvature* of the Universe as described by Einstein's relativity, but relativity does not say anything about the *topology* of the Universe.

Geometry describes the curvature of the Universe, and you can measure the geometry at any individual point. Topology, on the other hand, requires looking at the Universe as a whole. Topology describes the way different parts of the Universe are linked to one another. An example with which many will be familiar comes from video games. In some, if you exit the right side of the screen, you reenter from the left at the same height, and if you exit from the bottom, you reenter from the top. The geometry of the screen is flat, but the topology is the same as the surface of a bagel or donut: consider matching the right side with the left to give a tube, and then matching the top of the tube with the bottom, as in Figure 28.

This condition cannot be seen in a video game without bending your monitor, so the screen and the donut are not exactly the same, but they have the

Figure 28. A piece of paper, or a flat video game screen, has the same topology as the surface of a donut or a torus.

same topology—technically known as a torus—with different curvature. This example of a video game displayed on a screen is in two dimensions, but now consider the interior of a cube, where now we go out the top and reenter from the bottom, go out the left and reenter from the right, and go out the front and reenter from the back. This is the three-dimensional torus and is a perfectly valid possibility for an expanding Universe.[13] Indeed, the interaction of the randomness of quantum mechanics with the geometry of Einstein's relativity is thought by some to make an interesting topology more likely than not.

All this exiting and reentering means that if we look far enough in one direction, we would be able to see back to ourselves: we would expect the Universe in each direction to look like copies of itself, again and again. But the finite speed of light and the finite age of the Universe make this more complicated: as we look further away, we would see copies of the Universe as it was long ago, not today. And if that cube is large compared to the distance to the last scattering of the CMB, we would not see any copies at all: the nearest would be farther than light could have traveled since the Big Bang.

Even in this case, there might still be weaker visible effects: the Universe is no longer the same in all directions (isotropic). Now, that fundamental cube picks out some directions as special, which could then have some influence on the expected patterns in the CMB. Could this be responsible for the "anomalies" observed in the CMB? Frankly, I doubt it, but the possibility is too exciting to ignore, and I've spent some of my spare scientific time over the last couple of decades trying to work out ways to test it.[14]

Beyond the Hubble tension and CMB anomalies, there are other observations that might point to problems with our standard ΛCDM model. Results from CMB experiments like Planck can make predictions for the distribution of structure in the present-day Universe, which we can then compare to other experiments that look more directly at the distribution of matter today.

One promising method for observing that distribution is through its effects on the path of light rays coming from distant galaxies. The intervening matter will bend those rays—through the gravitational lensing predicted by Einstein's theories—and distort the shapes of the distant galaxies, allowing us to get a census of matter in the local Universe (galaxies relatively near the Milky Way). As with the Hubble constant, these very different measurements of the same quantity give results that are quite different from each other, compared to the error bars. If those are correct, we would expect at least this level of disagreement considerably less than one percent of the time.

Another example: with the launch of the recent James Webb Space Telescope, we can observe galaxies that are farther away, and hence earlier in time, than ever before. We are starting to find very large galaxies forming less than half a billion years after the start of the expansion of the Universe. Was there enough time to form these galaxies (and everything in them) so quickly? Some initial calculations predict that such massive galaxies should be very rare in these early times, so rare that we should not expect to see any of them.[15]

I could go on listing possible discrepancies between the predictions of the model and various observations, and I am certain that more will be found: the job of a scientist is to come up with theories and new ways to test those theories.

THIRTEEN

The Limits of Knowledge

If you take the literature at its word, some things
are so improbable, they're practically assured.
— The Paranoid Style, "I'd Bet My Land and Titles"

All this constant poking and prodding and testing with the singular goal of falsifying a scientific model means that we must take care in interpreting the results of any such tests. If you examine *anything* closely enough, you will find discrepancies with your expectations.

Consider, instead of cosmology, the assassination of John F. Kennedy. The few minutes around 12:30 p.m. on November 22, 1963, and the few hundred feet around Dealey Plaza in Dallas, Texas, have been examined as closely as any other event and place in human history. There are a small number of photographs and films, along with forensic evidence and eyewitness testimony, all of which can be compared to our model of what happened that day.

Everything that happens over the history of the Universe is ultimately governed both by the laws of physics and the random initial conditions present after a presumed period of cosmic inflation. The same is true of what happened around 12:30 p.m. on November 22, 1963: the laws of physics and biology and sociology governed what happened that day, but our ignorance of all the details leading up to the assassination, along with the inevitable randomness of messy human events, gets in the way of any ironclad predictions. Hence the persistence of conspiracy theories involving a second gunman, magic bullets, and inconsistencies in records such as the infamous Zapruder film or the supposedly definitive US government's *Warren Report* on the assassination, which supports the idea that the assassination was carried out by a lone gunman, Lee Harvey Oswald. But if you look at thousands of things, you will find many one-in-a-thousand coincidences.

So if you go looking for seeming contradictions (that is, for unlikely—but not impossible—events), you will find them. But, in the case of Kennedy's assassination, that does not (necessarily) prove that the lone-gunman theory is wrong. A technical term for this is *a posteriori reasoning*, but a better phrase that often applies is the "look elsewhere effect." If you are looking for some interesting effect over a very wide range of possibilities, to correctly assign the probability of finding some sort of anomaly, you must account for all the places that you looked but did not find one. In a vast and random Universe, even one governed by strict laws of physics, unlikely events are inevitable.

Our observations of the Universe are necessarily local: the finite speed of light gives what the laws of physics tell us is an absolute speed limit for the transmission of any information at all: we can learn about events one second ago only within a sphere of radius of one light-second, events one hour ago within a radius of one light-hour, one year ago within one light-year, and so on up through the ultimate limit of the age of the Universe and its related distance. This is the *light cone* that I mentioned earlier, closely related to the concept of *horizon*, which arose in the problems solved by inflation.

We can never get any information from outside of the horizon delineated by our light cone. This remains true even when accounting for the expansion of the Universe, including the possibility of accelerated expansion due to early inflation or more recent dark energy.[1]

Even though this region will grow with time, it will always be finite. Our best theories, therefore, can only be tested on the information from a limited region of the Universe. Can we be sure that our laws of physics, written down in our textbooks, apply beyond the horizon?

This circumscription of our observations to a finite region makes it hard to even define our Universe. Should we be content with a set of laws that describe only all that is within this limited purview? Usually, physicists hope for something grander, a set of laws that describe everything, everywhere, forever. So we are forced to ask what lies beyond the horizon, outside of our light cone, in places that we will never see, and from where we can never get any information.

Our model of inflation gives us some insight into how we might start to understand the status of the part of the Universe that we can observe. In this model, a small volume, somewhat larger than we can observe today, inflated to become flat and nearly uniform. Perhaps that is all we can say.

But the very fact of inflation's likely occurrence opens further possibilities: if inflation happened to the region destined to become our observable Universe, then, obviously, inflation is allowed by the laws of physics. Perhaps, then, inflation happened elsewhere, too. What does that mean?

Imagine the Universe before inflation as an ocean, burbling and roiling, particles and fields fluctuating within the randomness allowed and demanded by quantum mechanics. Sometimes, a tiny region will calm and achieve the conditions needed for inflation to occur. Suddenly, this volume will begin its rapid expansion. This newly independent region will become its own universe, losing its connection to what used to be nearby.

This process is called chaotic, or eternal, inflation, an idea that grew in the late 1980s out of the new inflation theories of Linde, Albrecht, and Steinhardt, along with Alexander Vilenkin.[2] This reaffirmed the idea that inflation erases information about the early Universe, funneling any region that manages to inflate toward a flat and uniform state.

Careful calculations show that those regions usually remain disconnected from each other. With inflation itself erasing details of the initial conditions of the patch, and without any way to communicate between them, we may as well call any such inflating region its own universe.

In some theories, the possibilities are weirder still. We don't know what the ultimate theory that connects gravity and particle physics will look like. Some speculation points toward the idea of *string theory*. Before the 1970s and 1980s, physicists thought that the fundamental constituents of the Universe would be point-like particles acting under the influence of quantum mechanics. This manifested itself as "quantum field theory," which successfully described all the interactions that describe chemical interactions, the structure of atoms, and the details of nuclear physics. But quantum field theory is complicated and may not even be mathematically consistent— many calculations result in infinite numbers that don't make a lot of sense at first. These can be "regularized" and "renormalized" using even more beautiful and complicated mathematics, but still leave us with the hint that our theory has swept something important under the rug.

String theory took up these hints, starting with the idea that the fundamental constituents of the Universe are not point-like (zero-dimensional) particles, but are instead one-dimensional—colloquially called "strings"—that still behave as dictated by the rules of quantum mechanics. It was quickly realized that this string theory has two remarkable advantages over the theory of point particles. First, the infinities of field theory are no longer present. Second, it very naturally predicts a force that behaves just like gravity. So string theory very quickly became a candidate for reconciling gravity and quantum mechanics.

Over the years, string theory has proven to be a complicated theory, both mathematically and physically. In the 1990s, string theory was subsumed into something called "M-theory" (where "M" can stand variously for "magic," "matrix," or "membrane"). The details of M-theory have not been worked out, but an important feature is that it can only make sense if there are a total of 10 or 11 dimensions.

But we don't live in a 10- or 11-dimensional world; somehow the theory also needs to explain how we only experience three-dimensional space and one dimension of time. If this theory is correct, those other "extra dimensions" need to somehow be made small—so small that we don't notice them. Imagine two-dimensional creatures constrained to live between two adjacent plates of glass. If the distance between the plates is very small compared to the size of the creatures, they would not notice that extra third dimension. If the remaining dimensions were similarly *compactified*, we too would think that we live in a three-dimensional universe.

In these theories, the details of this quantum mechanical compactification make a big difference—it not only determines the number of large dimensions, but also the properties of the particles and fields that exist in those dimensions. However, there are a lot of ways for this compactification to happen. Really, a lot: at least 10^{200} (1 followed by 200 zeros, which is the square of the slightly better known "googol": 10^{100}) but possibly many, many more.

So, perhaps the model of particle physics that describes the Universe is not a fundamental consequence of our theories, but as much an accident as the number of stars in the sky, or the locations of the molecules in the air of the room in which I'm writing these words, or the different varieties of frogs and toads that have evolved on Earth.

The properties of the resultant universes as the details of the compactification vary is called the "string landscape." And we still don't know whether any of those ways give us a universe like the one we live in, with three dimensions of space, as well as quarks, electrons, neutrinos, photons, gravity, dark matter, and dark energy.

And even if a universe like ours does exist on this vast landscape, why are we lucky enough to live there? Why did the Universe turn out like this? Most universes would not be conducive to life, certainly not to "life as we know it." Is it just good luck that the Universe supports life?

The idea of eternal inflation provides a possible answer. The roiling sea now takes place in 10 or 11 dimensions, and the randomness of quantum mechanics corresponds to different compactifications. Some of the locations on the landscape have particles that behave like an inflaton and get into just the right state to start an epoch of inflation.

Now, we have a way of producing universes, randomly selected from the landscape of possibilities.

But let's be more careful with our language. An inflating patch may have some of the properties usually ascribed to a "universe," but it categorically lacks what may be its defining characteristic, according to the *Oxford English Dictionary*:

> Universe: all existing matter and space considered as a whole; the cosmos.

This single inflating patch certainly isn't "all existing matter and space" (although we might need a noncircular definition of "cosmos" to fix the meaning here). Nonetheless, physicists have taken on a new term to describe the collection of individual inflating universes, along with the regions that have not yet undergone inflation: the *multiverse*.

I first encountered the concept in comic books and science fiction from the 1960s and 1970s, a term that referred to worlds with similar properties but different details: an evil Terran Empire instead of the beneficent Federation of Planets in *Star Trek*'s "Mirror, Mirror" episode, or the plethora of superhero-filled worlds with their different versions of Superman and the Flash.

In fact, the term was first used by the American philosopher and psychologist William James: "Visible nature is all plasticity and indifference, a moral multiverse, as one might call it, and not a moral universe."[3] Despite the different context, James's coinage works for us, emphasizing at least the possibility of radical differences across the *multi*verse, in contrast to the implied sameness of a *uni*verse. Only in the 1990s did "multiverse" become common in the scientific literature.[4]

The multiverse of eternal inflation demands a plethora of universes, possibly vastly different from one another. This may be reminiscent of our discussion of the many-worlds interpretation of quantum mechanics, and indeed the collection of quantum worlds has also been dubbed a version of the multiverse.[5]

These differing multiverses suggest a radical unification: Perhaps the many worlds and inflating regions are the same? If the inflationary multiverse is big enough, anything that can happen will happen, eventually, just as in the branches of the quantum mechanical wave function.

The Anthropic Idea

It is daunting to be confronted by such a vast multiverse. Copernicus and Galileo moved Earth from the center of creation. Astronomers of the eighteenth and nineteenth centuries showed that our solar system is just one among many in the Milky Way, and Hubble showed that the Milky Way is one of innumerable galaxies. And now our entire Universe might just be a random fluctuation.

Aside from enforcing a radical humility about our own importance, this model of an inflationary multiverse allows us to do what we always do with a model—make inferences using Bayes' theorem. We have a model, but what should we use as our data? We could simply require a detailed match with all our observations: the standard model of particle physics, the ΛCDM cosmology, the existence of Earth and our solar system, the Milky Way, the observed distribution of galaxies and the pattern of the CMB.

But perhaps this is too specific. In a random multiverse, we might want to know not only that our Universe is possible, but whether it is somehow likely. So rather than condition upon the fine-grained details of our observed Universe, perhaps we can step back and condition upon something broader: the existence of intelligent life, say. But now we have a problem not of physics, but of biology. We don't know what processes underlie the evolution of life at all, much less intelligence and culture. Does all life require water? How, and how often, do complex multicellular creatures—presumably required for intelligence—evolve from simpler forms of life? How do intelligence and culture arise?

The idea of quantifying these questions in order to work out the probabilities about the Universe is known as the *anthropic principle*, coined by Brandon Carter in 1973, although Robert Dicke, one of the scientists behind the discovery and interpretation of the cosmic microwave background, used similar reasoning in the 1950s.[6] Both Carter and Dicke wondered why the Universe seemed to be just right for life (as we know it). For example, we should expect to find life only during a period late enough for stars and planets to have formed, but not so late that most of the atoms in the Universe would already have been processed through stars. Dicke and Carter argued that we live at a special time, just when the Universe allows the kind of complicated structures required for life.

Others have used this anthropic principle to make stronger claims, not just about when life could arise, but about the properties of the Universe itself. Steven Weinberg noted that the most physically natural value for the cosmological constant, Λ, is to be very, very large compared to our observations. However, if Λ were much larger than its observed value, the Universe would have undergone a second epoch of inflation even before the formation of stars, and hence would not allow life.[7]

Similar arguments have been used at an even lower level to argue that all the fundamental parameters of physics are constrained by the observation that carbon-based, intelligent life exists. In the 1950s, Fred Hoyle pointed out that the properties of the carbon nucleus need to be tuned very precisely for stars to form and enable the thermonuclear fusion of hydrogen first to helium and then to carbon, one of the necessary elements for life (indeed, the presence of carbon is defined to be the marker of an "organic" molecule). But these properties depend in a very complicated way on the properties of quarks and other fundamental particles—there is no simple underlying physical principle to ensure this property of carbon.[8]

More recently, anthropic arguments have been used to justify our place in the string landscape, or in the outcome of any scenario with eternal inflation: we would not be here to talk about the string landscape if we did not end up in a universe that looks like ours.

But this seems like a tautology. We use Bayes' theorem to measure the values of the constants that govern our Universe; can we also use it to determine the probability of the formation of the Universe itself? All we have to do is return to our slogan: all probabilities are conditional. Everything depends on the model—everything, including the way we interpret our data, and even what data we choose to consider.

When we are just measuring parameters from scientific experiments, this is often straightforward, as the data we have gathered overwhelm whatever prior probability that I might assign. But when considering such fuzzy data as the existence of life (or sentient life, or a sufficiently richly complicated Universe for the formation of life, or . . .) we have to specify our model extremely carefully for the probabilities to be meaningful. And then we must always ask of our models, "Compared to what?" I need to assign a prior to each possible universe, but absent a compelling theory, this makes little sense.

Does the string landscape, with or without the idea of chaotic inflation to actually realize multiple possible universes, give any guidance for how to assign a prior over the 10^{200} possible universes? As of now, there are some very tentative ideas for why certain regions of the landscape might be special, but no real principles for choosing some possible universes over others beyond the self-evident fact of our own existence—a measurement of the properties of the Universe, rather than a prediction.

For this reason, many scientists are skeptical of attempts to use anthropic reasoning to understand a scientific theory. In their 1989 textbook *The Early Universe*, Mike Turner and Rocky Kolb of the University of Chicago (full disclosure: I was a graduate student at Chicago studying cosmology at the time this textbook was coming together, and Turner was my co-supervisor) briefly discussed the anthropic principle, with this footnote: "It is unclear to one of the authors how a concept as lame as the 'anthropic idea' was ever elevated to the status of a principle."[9]

Others have labeled anthropic arguments as inherently unscientific, based on a fact that is important to us—we are here!—but of no import to what we usually take to be physical laws indifferent to our existence. Some feel the multiverse itself to be unscientific, untestable almost by definition, with all the other universes inaccessible. This attitude is an outgrowth of the operational nature of the revolutions of twentieth-century physics, which ended up with both quantum mechanics and general relativity, defined in terms of the measurements that can be performed on a system, rather than the state of the system itself. In this extreme positivist view, if something can't be observed, it is not in the realm of science.

This seems a very limiting restriction: things can be *true* even if they are not directly testable. It very well could be the case that M-theory, or something like it, a theory that only vaguely constrains the possible state of our Universe without determining its properties in detail, is true.

Certainly, we would prefer an alternative theory that predicted exactly one universe, with exactly the properties of ours. Imagine that we encode our prior understanding of M-theory by using the principle of indifference, assigning the same probability to each of its 10^{200} possible universes. Conversely, the theory that directly predicts our Universe has no free parameters. This means that the data prefer this theory by that same factor of 10^{200}; unless our theoretical prejudice— our prior preference for M-theory compared to the "just-so" alternative—can overcome those steep odds, we should strongly prefer the simpler one.

Bayes' theorem tells us that the occurrence of an unlikely event, such as the creation of a universe with particular physical laws, does not give any information at all about the existence of other universes. If those other universes don't have any observational repercussions, we cannot get any information beyond our priors.[10]

But we should never discount the progress of science and technology: just because a theory does not seem testable today, it will not necessarily remain so, as attested by the successful measurements of the incredibly small fluctuations in the cosmic microwave background, and the even smaller disturbances due to gravitational radiation observed by the LIGO experiment.

Anything Is Possible

Is this a disappointing state of affairs, confronted with a theory that might not be able to predict the laws of our Universe? Are we disappointed that Darwin's theory of evolution does not predict the existence of frogs, of cats, of intelligent humans? No, because we see evolution as an algorithm, a fully scientific theory describing a process rather than an outcome. Perhaps that might be all we can ask of our theories of physics, as well: a consequence of a small set of physical principles that play out in the multiverse. Is that enough for us to be certain about the laws of the Universe?

In some important sense, this can't possibly be the case. One very boring set of physical laws that cannot possibly be disproven is just a list of the trajectories of all the particles in the Universe, from the dawn of time until now. The Universe is just one damned thing after another. This is a terrible theory—it perfectly describes what happens, without any attempt at explaining why or at predicting the future. And it is incredibly prolix: it could not even be stored or written down within the universe that it describes—since it would take more than one particle's worth of information to describe each particle! So perhaps this is not a realistic theory.

But other insidious and untestable theories abound. Descartes's seventeenth-century skepticism led him to consider the possibility of a malignant demon, controlling our senses and fooling us into thinking that the real world is very different from that which we perceive. All Descartes could be sure of was, of course, his famous aphorism "I think, therefore I am." Modern versions of Descartes's demon are the computational overlords of *The Matrix* films, projecting a virtual world into the brains and minds of most of the human population.

Even more bizarre—if only because it is somewhat seriously considered by scientists—is the possibility of what are known as "Boltzmann brains," named after the physicist who was among the first to understand the statistical properties of thermodynamic randomness. We learned that tiny quantum mechanical fluctuations at the end of inflation eventually grow to become all the structure that we observe in the Universe today. More broadly, quantum mechanics tells us that random assemblages of whatever particles and fields are allowed by the underlying laws of physics will sometimes just (for lack of a good technical term) pop into existence. There are constraints on this process: energy must be conserved, and the larger and more complicated an object is, the less likely it is to pop. But all structures are possible, not just the small smooth lumps that we infer from our observations of the cosmic microwave background and the current pattern of galaxies.

In a sufficiently large universe, anything may be created. This includes, say, a human brain in exactly the state that I believe mine to be in right now, with 50-plus years of memories of life on a planet called Earth with my family and friends, as I sit writing this book. Except that those memories and experiences will be completely false. Boltzmann brains are exceedingly rare. However, a universe (or multiverse) undergoing eternal inflation has a lot of time available. And, in some such scenarios, Boltzmann brains could be more common than real brains, inhabiting an inflating universe that itself only came into existence because of a much larger—and hence even less likely—quantum fluctuation. So if all you have is Descartes's slogan "I think, therefore I am," the "I am" might be more likely to be a Boltzmann brain than a real one in the real world in some models. Descartes used the existence of a benevolent god to ensure the reality of his perceptions; it's harder for us with just the laws of physics to rely on to ensure the reality of ours.[11]

But can we even rely on those laws? Can we rely on there even being a single set of laws that describes all creation, be it a multiverse or a universe? David Lewis, a late twentieth-century philosopher, described a stance he called *modal realism*, which boils down to the idea that any possible universe (although he preferred the term "world") is a real universe: if you can imagine it, it exists. But "exists" is a slippery term in this context: we need to treat this imagined universe as having its own independent existence, but it is completely independent of and inaccessible to the world in which we (happen to) live. Our job is still to work out which universe is our own.[12]

A perhaps more precise version of this idea makes the concept of a possible universe more specific: any self-consistent set of mathematical equations describing physical laws comes with its own universe realizing those laws.[13]

These seemingly radical ideas take us back to the beginnings of philosophy. Pythagoras is thought to have believed that the world consists only of numbers; Plato wrote of the reality of the ideal forms of which the concepts and objects in our world are somehow just pale imitations. I prefer Hume's more modern take: we cannot derive "ought" from "is." That is, the real world can never give us enough information for us to be sure of the laws that govern it.

We seem to be back where we started, wanting to go from some finite amount of data to universal principles—Hume's problem of induction, which led us to models, probability, and Bayes' theorem, tools that we can apply even in the face of uncertainty.

To me, this is a liberating prospect: discovering a new model for the world, or even making small progress refining the models that are already out there, is a human-sized act of creation and creativity, the best way to understand the reality of the world around us.

NOTES

Chapter 1. We Are Scientists

1. R. Owen, *Where Did All the Dinosaurs Go? (Little Scientists BIG Questions)* (Ruby Tuesday Books, 2018).
2. Perhaps a bad example, since I've heard every single one of the Beatles' songs many times.
3. J. Watson and F. Crick, "Molecular Structure of Nucleic Acids: A Structure for Deoxyribose Nucleic Acid," *Nature* 171 (1953): 737–738, doi:10.1038/171737a0; M. Wilkins, A. Stokes, and H. Wilson, "Molecular Structure of Nucleic Acids: Molecular Structure of Deoxypentose Nucleic Acids," *Nature* 171 (1953): 738–740, doi:10.1038/171738a0; R. Franklin and R. Gosling, "Molecular Configuration in Sodium Thymonucleate," *Nature* 171 (1953): 740–741, doi:10.1038/171740a0.
4. M. J. Tobin, "April 25, 1953: Three Papers, Three Lessons," *American Journal of Respiratory and Critical Care Medicine* 167 (2003): 8, 1047–1049, doi:10.1164/rccm.2302011.
5. A. Eddington, *The Expanding Universe* (Pelican, 1940).
6. C. Will, "Henry Cavendish, Johann von Soldner, and the Deflection of Light," *American Journal of Physics* 56 (1988): 413, doi: 10.1119/1.15622.
7. P. G. Ferreira, *The Perfect Theory: A Century of Geniuses and the Battle over General Relativity* (Little, Brown, 2014); C. M. Will, "The Confrontation Between General Relativity and Experiment," *Living Reviews of Relativity* 17, 1 (2014): 4, doi:10.12942/lrr-2014-4.
8. F. W. Dyson, A. S. Eddington, and C. Davidson, "A Determination of the Deflection of Light by the Sun's Gravitational Field, from Observations Made at the Total Eclipse of May 29, 1919," *Philosophical Transactions of the Royal Society of London, Series A* 220 (1920): 291–333. The paper reports their error bars as a "probable error" of 0.12 arc seconds, which translates to what would today be reported as 0.18 arc seconds.

9. Eddington, *The Expanding Universe.*
10. Quoted in D. Oriti, "Tightrope-Walking Rationality in Action: Feyerabendian Insights for the Foundations of Quantum Mechanics," *International Studies in the Philosophy of Science* (2024): 1–33, doi:10.1080/02698595.2024.2351917.
11. "Data" is the plural of "datum."
12. The perceptive will note that the very idea of object permanence is a scientific model for human development!

Chapter 2. The Problem and the Promise of Induction

1. H. Putnam, *Reason, Truth and History* (Cambridge University Press, 1982).
2. Well, not entirely. We also observe gravitational waves, which also move at the speed of light, so the argument is the same. For nearby objects, we sometimes get information from massive objects moving somewhat slower than the speed of light, so our knowledge of the nearby Universe can be a little more complete.
3. F. Ramsey, "Truth and Probability," in *The Foundations of Mathematics and Other Logical Essays*, edited by R. B. Braithwaite (Harcourt, Brace, 1931), 156–198.
4. P. A. Schilpp, ed., *Albert Einstein: Philosopher-Scientist* (Cambridge University Press, 1949).
5. W. V. O. Quine, "Epistemology Naturalized," in *Ontological Relativity and Other Essays* (Columbia University Press, 1969), 69–90, doi:10.7312/quin92204–004.

Chapter 3. What We Talk About When We Talk About Probability

1. There is debate about the meaning of "planet" even in our own solar system, as exemplified by the recent expulsion of Pluto.
2. Quoted in M. Banks, "The Daily Show Does CERN," Physics World, May 5 (2009), https://physicsworld.com/a/the-daily-show-does-cern/.
3. Ó. B. de Finetti, *Theory of Probability* (Wiley, 1974).
4. This same slogan was expounded by David Lindley in L. J. Savage et al., *The Foundations of Statistical Inference* (Methuen, 1962), 83.
5. I hope that you believe me when I say that I have exaggerated the probability of my forgetting my keys in all these cases.
6. For Cox, see R. T. Cox, "Probability, Frequency and Reasonable Expectation," *American Journal of Physics* 14, 1 (1946): 1–10, doi:10.1119/1.1990764. For Jaynes, see E. T. Jaynes, *Probability Theory: The Logic of Science* (Cambridge University Press, 2002). For van Horn, see K. van Horn, "Constructing a Logic of Plausible Inference: A Guide to Cox's Theorem," *International Journal of Approximate Reasoning* 34, 1 (2003): 3–24, doi:10.1016/S0888–613X(03)00051–3.
7. Thanks to my colleague David Hogg for pointing out the etymology of the term to me.

8. Variants of this argument that rely less strongly on betting odds have also been put forward—for instance in Ramsey, "Truth and Probability"; de Finetti, *Theory of Probability*; J. Barnardo and A. Smith, *Bayesian Theory* (Wiley, 1994).
9. "In the long run, we are all dead"—J. M. Keynes, *The Tract on Monetary Reform* (Macmillan, 1923). Keynes did write about probability, but this quotation is about economics.

Chapter 4. "Our Deceased Friend Mr Bayes"

1. Mathematically, the equals sign should really be translated as "proportional to."
2. D. A. Mistry, J. Y. Wang, M. E. Moeser, et al., "A Systematic Review of the Sensitivity and Specificity of Lateral Flow Devices in the Detection of SARS-CoV-2," *BMC Infect Dis* 21 (2021): 828, doi:10.1186/s12879-021-06528-3. The World Health Organization and the European Commission define acceptable performance as a minimum 80 percent sensitivity and a maximum 2 percent false positive rate—see https://www.tga.gov.au/products/covid-19/covid-19-tests/covid-19-rapid-antigen-self-tests-home-use/covid-19-rapid-antigen-self-tests-are-approved-australia.
3. The clarity of an expression like this is why scientists like to use mathematical notation (even when we worry that it might scare away some of our readers)!
4. M. Gladwell, *Outliers: The Story of Success* (Little, Brown, 2008).
5. As Gladwell says, no amount of practice will make an innately slow runner fast. And at least in some fields such as classical music, it's very helpful to have both talent and great teachers. Gladwell, "Complexity and the 10,000 Hour Rule," *New Yorker*, August 21, 2013; K. A. Ericsson, R. Krampe, and C. Tesch-Romer, "The Role of Deliberate Practice in the Acquisition of Performance," *Psychological Review* 100, 3 (1993): 363–406.
6. Readers well-versed in the mathematics of probability will notice that I am playing somewhat fast and loose and not distinguishing between the concept of *probability* and *probability density*, where the latter is appropriate for these measurements where I am allowing the quantity that I am measuring to take on any real number value at all.
7. "Jerusalem" is taken from a longer work, "Milton," adapted and set to music by Hubert Parry only after the First World War.
8. I grew up in the United States, where something is old if it was built a century ago and students are always amused to hear that "modern" history starts in 1789. Since I have lived in Britain for the last 20 years, a country where the eighteenth century sometimes seems recent, this makes a lot more sense, as does the quote from Chinese Premier Zhou Enlai in the early 1970s that "it's too early to tell" the effects of the French Revolution (though he was probably referring to the 1968 student riots).
9. The artist Eduardo Paolozzi has adapted the work, showing a more explicitly heroic Newton in a giant sculpture outside the British Library, an homage to the world-views of both Blake and Newton echoed in the pages inside.

10. S. M. Stigler, "Thomas Bayes's Bayesian Inference," *Journal of the Royal Statistical Society: Series A (General)* 145 (1982): 250–258, doi:10.2307/2981538.
11. J. Davies, *History of Wales* (Penguin, 1990), 32–35.
12. S. M. Stigler, "Richard Price, the First Bayesian," *Statistical Science* 33 (2018): 117–125, doi:10.1214/17-STS635.
13. D. Hume, *A Treatise of Human Nature*, book 1, part 4, section 1 (1739).
14. S. M. Stigler, "The True Title of Bayes' Essay," *Statistical Science* 28, 3 (2013): 283–288, doi:10.1214/13-STS438; Stigler, private communication with the author, 2022.

Chapter 5. Models and Measurements

1. They Might Be Giants, "Why Does the Sun Really Shine? (The Sun Is a Miasma of Incandescent Plasma)," a song from the record *Here Comes Science*, responding to the song "Why Does the Sun Shine? (The Sun Is a Mass of Incandescent Gas)," by Hy Zaret, also recorded by They Might Be Giants.
2. Tipler has argued that Kelvin might have been able to rescue his model had he considered a model with something almost as dense at the center of the sun, a possibility that his nineteenth-century knowledge could not have definitively ruled out. F. J. Tipler, "Perry, Kelvin, and the Age of the Sun," *European Physical Journal* H 38 (2013), doi:10.1140/epjh/e2012–30021–5.
3. This fact makes it particularly dangerous to use them as the canonical examples of scientific theories.
4. O. Römer, "Demonstration touchant le mouvement de la lumiere trouvé par M. Römer de l'Academie Royale des Sciences," December 7, 1676.
5. For completeness, the different particles are the up, down, charm, strange, top, and bottom quarks; the electron, muon, tau leptons; and their three partner neutrinos, the photon, W, Z, and gluon gauge, and the Higgs bosons.
6. It also inspired one of the great mathematical astrophysicists of the twentieth century, Subrahmanyan Chandrasekhar, whom we will meet in the next chapter, to translate the book anew, not into a modern language, but into modern mathematics. S. Chandrasekhar, *Newton's Principia for the Common Reader* (Clarendon, 2003).
7. E. P. Wigner, "The Unreasonable Effectiveness of Mathematics in the Natural Sciences. Richard Courant Lecture in Mathematical Sciences Delivered at New York University, May 11, 1959," *Communications on Pure and Applied Mathematics* 13, 1 (1960): 1–14, doi:10.1002/cpa.3160130102.
8. This is a simplification and variation of an argument by the philosophers Quine and Putnam called the "indispensability argument": numbers, and the relation between them, are "indispensable" to a scientific theory and so must be real in some sense. H. Putnam, "What Is Mathematical Truth," *Historia Mathematica* 2 (1975): 529–533.

9. A. Baker, "Are There Genuine Mathematical Explanations of Physical Phenomena?," *Mind* 114, 454 (2005): 223–238; J. Yoshimura, "The Evolutionary Origins of Periodic Cicadas During Ice Ages," *American Naturalist* 149, 1 (1997): 112–124.

10. A. Einstein, "On the Electrodynamics of Moving Bodies," *Annalen der Physik* 17 (1905): 891–921.

11. I am indebted to Daniel Dennett for this way of at looking evolution. See D. C. Dennett, *Darwin's Dangerous Idea: Evolution and the Meanings of Life* (Simon & Schuster, 1995).

12. Although Box is often cited as the source of this well-known quote, it appears he did not say it in so many words. See, e.g., G. E. P. Box, "Robustness in the Strategy of Scientific Model Building," in *Robustness in Statistics*, ed. R. L. Launer and G. N. Wilkinson (Academic Press, 1979), 201–236, doi:10.1016/B978-0-12-438150-6.50018-2.

13. E. Thompson, *Escape from Model Land: How Mathematical Models Can Lead Us Astray and What We Can Do About It* (Basic Books, 2022).

14. I attended a conference in the early 2000s with the eminent Bayesian statistician Jim Berger, who (paraphrasing from a long-ago memory) quipped that all of statistics is about averaging.

15. N. Silver, *The Signal and the Noise: Why So Many Predictions Fail—but Some Don't* (Penguin Publishing Group, 2012).

16. N. Silver, "Why FiveThirtyEight Gave Trump a Better Chance Than Almost Anyone Else," FiveThirtyEight, November 11, 2016, https://fivethirtyeight.com/features/why-fivethirtyeight-gave-trump-a-better-chance-than-almost-anyone-else/.

Chapter 6. Physicists and Philosophers

1. "Meeting of the Royal Astronomical Society, Friday 1935 January 11," *The Observatory* 58, 729 (1935): 38.

2. A. Eddington, "On Relativistic Degeneracy," *Monthly Notices of the Royal Astronomical Society* 95, 3 (1935): 194.

3. A. Miller, *Empire of the Stars: Obsession, Friendship, and Betrayal in the Quest for Black Holes* (Mariner Books, 2005).

4. S. Chandrasekhar, quoted in K. Wali, "Chandrasekhar vs. Eddington—an Unanticipated Confrontation," *Physics Today* 35, 10 (1982): 33–40.

5. P. K. Feyerabend, *Against Method*, 4th ed. (Verso Books, 2010).

6. Lakatos had intended to write a comprehensive response to Feyerabend's ideas, but, alas, died in 1974 with no substantial work done, although we now have access to the pair's correspondence over the years to give a flavor of what might have appeared. I. Lakatos, P. Feyerabend, and M. Motterlini, *For and Against Method* (University of Chicago Press, 1999), doi:10.7208/9780226467030.

7. Feyerabend, *Against Method*, vii.

8. Feyerabend, *Against Method*, 21.

Chapter 7. Entropy and Information

1. As described in A. Berdondini, "Application of the von Mises' Axiom of Randomness on the Forecasts Concerning the Dynamics of a Non-Stationary System Described by a Numerical Sequence," *SSRN Electronic Journal* (2019), doi: 10.2139/ssrn.3319864.
2. At least to a very good approximation. In fact, some of the very fine details might depend on quantum mechanics, which, as we will see, is actually unpredictable.
3. C. E. Shannon, "A Mathematical Theory of Communication," *Bell System Technical Journal* 27 (1948): 379–423, doi:10.1002/j.1538–7305.1948.tb01338.x.
4. For example, see https://rosettacode.org/wiki/Markov_chain_text_generator.
5. OpenAI, ChatGPT (Large language model) (2023), https://chat.openai.com/chat.
6. US Library of Congress, "Fascinating Facts," https://www.loc.gov/about/fascinating-facts/.

Chapter 8. Entropy and Matter

1. P. K. Dick, *Do Androids Dream of Electric Sheep?*, S. F. Masterworks (Gateway, 2010).
2. M. Kondo, *The Life-Changing Magic of Tidying Up: The Japanese Art of Decluttering and Organizing* (Ten Speed Press, 2014).

Chapter 9. Quantum Randomness

1. In fact it would be somewhat more complicated due to the *magnetic* part of Maxwell's theory, but it still wouldn't match the observations of atoms.
2. Technical aside: in this case, the squared wave function gives something called a probability density; we really want to ask about the probability of finding the electron in some small but finite box near a position, which requires the further mathematical operation of integration.
3. Or dead.
4. For the thought experiment by EPR, see A. Einstein, B. Podolsky, and N. Rosen, "Can Quantum-Mechanical Description of Physical Reality Be Considered Complete?," *Physical Review* 47, 10 (1935): 777–780, doi:10.1103/PhysRev.47.777. For the version by Aharonov and Bohm, see D. Bohm and Y. Aharonov, "Discussion of Experimental Proof for the Paradox of Einstein, Rosen, and Podolsky," *Physical Review* 108, 4 (1957): 1070, doi:10.1103/PhysRev.108.1070.
5. J. von Neumann, in *Mathematical Foundations of Quantum Mechanics, New Edition*, ed. N. A. Wheeler (Princeton University Press, 2018).
6. G. Hermann, "Die naturphilosophischen Grundlagen der Quantenmechanik," *Naturwissenschaften* 23, 42 (1935): 718–721, doi:10.1007/BF01491142.

7. J. S. Bell, "On the Einstein Podolsky Rosen Paradox," *Physics Physique Физика* 1, 3 (1964): 195–200, doi:10.1103/PhysicsPhysiqueFizika.1.195.

8. From a 1947 letter to Max Born—"spooky action at a distance" is a common translation of the original German "spukhafte Fernwirkung."

9. S. Hossenfelder, "Superdeterminism: A Guide for the Perplexed," arXiv:2010.01324 (2020).

10. Often misattributed to Richard Feynman, this quip probably comes from M. D. Mermin: "Could Feynman Have Said This?," *Physics Today* 57, 5 (May 2004): 10–11, https://doi.org/10.1063/1.1768652.

11. R. Shankar, *Principles of Quantum Mechanics* (Plenum Press, 1994), doi:10.1007/978–1-4757–0576–8.

12. I. D. Toader, "Is Bohr's Correspondence Principle Just Hankel's Principle of Permanence?," *Studies in History and Philosophy of Science*, arXiv:2310.10192 (2023).

13. J. Honner, "The Transcendental Philosophy of Niels Bohr," *Studies in History and Philosophy of Science Part A* 13, 1 (1982): 1–29, doi:10.1016/0039–3681(82)90002–4.

14. I. Lakatos, "Falsification and the Methodology of Scientific Research Programmes," in *Criticism and the Growth of Knowledge*, ed. I. Lakatos and A. Musgrave (Cambridge University Press, 1970), 91–195.

15. Feyerabend, *Against Method*, 259–260.

16. S. M. Carroll and C. T. Sebens, "Many Worlds, the Born Rule, and Self-Locating Uncertainty," in *Quantum Theory: A Two-Time Success Story*, ed. D. Struppa and J. Tollaksen (Springer, 2014), doi:10.1007/978–88–470–5217–8_10.

17. L. Niven, "All the Myriad Ways," in *All the Myriad Ways* (Ballantine Books, 1971).

18. There is some evidence that Everett himself thought about this, but it has been discussed explicitly in the literature since the 1980s. E.g., M. Tegmark, "The Interpretation of Quantum Mechanics: Many Worlds or Many Words?," *Fortschritte der Physik* 46, 6–8 (1988): 855–862, doi:10.1002/(SICI)1521 3978(199811)46:6/8%3C855::AID-PROP855%3E3.0.CO;2-Q; D. Lewis, "How Many Lives Has Schrödinger's Cat?," *Australasian Journal of Philosophy* 82, 1 (2004): 3–22, doi:10.1080/713659799.

19. In the technical literature, this is called "the preferred basis problem."

20. W. H. Zurek, "Decoherence and the Transition from Quantum to Classical," *Physics Today* 44 (1991): 36–44 [abstract and updated version available online as "Decoherence and the Transition from Quantum to Classical—Revisited," arXiv:quant-ph/0306072 (2003).

21. M. Hemmo and O. Shenker, "The Preferred Basis Problem in the Many-Worlds Interpretation of Quantum Mechanics: Why Decoherence Does Not Solve It," *Synthese* 200, 261 (2022).

22. For example, C. A. Fuchs, N. D. Mermin, and R. Schack, "An Introduction to QBism with an Application to the Locality of Quantum Mechanics," *American Journal of Physics* 82, 8 (August 1, 2014): 749–754, doi:10.1119/1.4874855; C. A. Fuchs and B. C. Stacey, "QBism: Quantum Theory as a Hero's Handbook," arXiv:1612.07308 (2016); *Proceedings of the International School of Physics "Enrico Fermi": Course 197, Foundations of Quantum Theory* (2019).

23. The technical details of the QBist position are much more complicated. In particular, a full discussion of the mathematics of the QBist formalism would require us to consider not just the wave function, but its more complicated cousin, called the density matrix. Furthermore, although its main philosophical stance is completely general, QBism requires some as-yet-unproven mathematics for it to hold in its fullest and most elegant form. E.g., C. A. Fuchs, M. C. Hoang, and B. C. Stacey, "The SIC Question: History and State of Play," *Axioms* 6, 3 (2017): 21, doi:10.3390/axioms6030021.

24. Or, more properly, density matrices.

25. T. Palmer, "Superdeterminism Without Conspiracy," arXiv:2308.11262 (2023).

26. This is not strictly true. Some theories, such as de Broglie/Bohm's pilot-wave, have additional freedom that needs to be set up in such a way that the laws of ordinary quantum mechanics are obeyed. And both the pilot-wave theory and spontaneous collapse theory have not yet been convincingly cast in such a way that they can deal with the more elaborate version of quantum mechanics called "quantum field theory," which is needed to describe the details of the subatomic particle.

27. C. A. Fuchs, "Delirium Quantum or, Where I Will Take Quantum Mechanics If It Will Let Me," *AIP Conference Proceedings* 889, 1 (2007): 438–462, doi:10.1063/1.2713493.

Chapter 10. The Oldest Light

1. Now known as the Kunstinstitut Melly, then as the Witte de With Center for Contemporary Art.

2. T. Wright, "An Original Theory or New Hypothesis of the Universe" (1750).

3. I. Kant, *Universal Natural History and Theory of the Heavens or an Essay on the Constitution and the Mechanical Origin of the Entire Structure of the Universe Based on Newtonian Principles* (1755), https://publicdomainreview.org/collection/an-original-theory-or-new-hypothesis-of-the-universe/.

4. W. Herschel, "On the Construction of the Heavens," *Philosophical Transactions of the Royal Society of London* 75 (1785): 213–266.

5. M. A. Hoskin, "The 'Great Debate': What Really Happened," *Journal of Historical Astronomy* 7 (1976): 169–182.

6. A. van Maanen, "Preliminary Evidence of Internal Motion in the Spiral Nebula Messier 101," *Astrophysical Journal* (1916): xliv, 210–228.

7. V. Trimble, "H_0: The Incredible Shrinking Constant, 1925–1975," *Publications of the Astronomical Society of the Pacific* 108 (1996): 1073–1082, doi:10.1086/133837.

8. G. Lemaître, *Annales de la Société scientifique de Bruxelles* A (1927): 47, 49–59; E. P. Hubble, *Proceedings of the National Academy of Sciences* USA 15 (1929): 168–173. Lemaître's article was published in an obscure French-language journal and was translated into English only just after Hubble's results. See also M. Livio, "Mystery of the Missing Text Solved," *Nature* 479 (2011): 171; G. Lemaître, *Monthly Notices of the Royal Astronomical Society* 91 (1931): 483–490.

9. "IAU Members Vote to Recommend Renaming the Hubble Law as the Hubble-Lemaître Law," October 29, 2018, International Astronomical Union, https://www. iau.org/news/pressreleases/detail/iau1812/?lang.

10. R. Berendzen, "Geocentric to Heliocentric to Galactocentric to Acentric: The Continuing Assault to the Egocentric," *Vistas in Astronomy* 17 (1975): 65–83, doi:10.1016/0083–6656(75)90049–5.

11. Conversely, we know that things get cooler if we let them expand. This is exactly the principle on which refrigerators work as they expand (and then recompress) the liquids that circulate through them.

12. A. D. Sakharov, "Violation of CP Invariance, C Asymmetry, and Baryon Asymmetry of the Universe," *Journal of Experimental and Theoretical Physics Letters* 5 (1967): 24–27, doi:10.1070/PU1991v034n05ABEH002497.

13. The technical term is "CP Violation" and was discovered by Cronin and Fitch in 1964, for which they won the Nobel Prize in 1980. J. H. Christenson et al., "Evidence for the 2π Decay of the K_2^0 Meson System," *Physical Review Letters* 13, 4 (1964): 138, doi:10.1103/PhysRevLett.13.138.

14. See S. Weinberg's classic *The First Three Minutes*, an excellent, if somewhat technical, introduction to cosmology, concentrating on this particular event in the history of the Universe. S. Weinberg, *The First Three Minutes: A Modern View of the Origin of the Universe* (Basic Books, 1977).

15. R. A. Alpher, H. Bethe, and G. Gamow, "The Origin of Chemical Elements." *Physical Review* 73, 7 (1948): 803–804, doi:10.1103/PhysRev.73.803.

16. R. A. Alpher and R. Herman, "Evolution of the Universe," *Nature* 162 (1948): 774–775, doi:10.1038/162774b0.

17. A. G. Doroshkevich and I. D. Novikov, "Republication of: Mean Density of Radiation in the Metagalaxy and Certain Problems in Relativistic Cosmology," *General Relativity and Gravitation* 50 (2018): 138, doi:10.1007/s10714–018–2441-x.

18. A "white dielectric material," as Penzias later put it. D. Overbye, "Where the Universe Began," *New York Times*, September 4, 2023, https://www.nytimes .com/2023/09/04/science/astronomy-holmdel-antenna-microwaves.html.

19. P. J. E. Peebles, *Finding the Big Bang* (Cambridge University Press, 2009).

20. R. H. Dicke, P. J. E. Peebles, P. G. Roll, and D. T. Wilkinson, "Cosmic Black-Body Radiation," *Astrophysical Journal* 142 (1965): 414, doi:10.1086/148306; A. Penzias and R. Wilson, "A Measurement of Excess Antenna Temperature at 4080 Mc/s," *Astrophysical Journal* 142 (1965): 419, doi:10.1086/148307.

21. W. Sullivan, "Signals Imply a 'Big Bang' Universe," *New York Times*, May 21, 1965, https://www.nytimes.com/1965/05/21/archives/signals-imply-a-big-bang-universe .html.

22. E.g., "History of the CMB Dipole Anisotropy," Astro UCLA, https://www.astro .ucla.edu/~wright/CMB-dipole-history.html; E. K. Conklin, "Velocity of the Earth with Respect to the Cosmic Background Radiation," *Nature* 222, 5197 (1969): 971–972, doi:10.1038/222971a0; P. S. Henry, "Isotropy of the 3K Background," *Nature* 231, 5304 (1971): 516–518, doi:10.1038/231516a0; G. F. Smoot, M. V. Gorenstein, and R. A. Muller, "Detection of Anisotropy in the Cosmic Blackbody Radiation," *Physical Review Letters* 39, 14, APS (1977): 898–901, doi:10.1103/Phys RevLett.39.898.

23. A. H. Jaffe, A. Stebbins, and J. A. Frieman, "Minimal Microwave Anisotropy from Pertur-bations Induced at Late Times," *Astrophysical Journal* 420 (1994): 9, doi:10.1086/173538.

24. G. F. Smoot, "Structure in the COBE Differential Microwave Radiometer First-Year Maps," *Astrophysical Journal* 396 (1992): L1, doi:10.1086/186504.

25. Data from NASA's LAMBDA website, https://lambda.gsfc.nasa.gov/. Plotted using the Healpix software, described in K. M. Gorski et al., "HEALPix: A Framework for High-Resolution Discretization and Fast Analysis of Data Distributed on the Sphere," *Astrophysical Journal* 622 (2005): 759–771, doi:10.1086/427976.

26. Based on Z. Zonca, https://www.zonca.dev/posts/2013–08–08-healpix-map-of-earth-using-healpy. Plotted using the Healpix software (Gorski et al., "HEALPix.").

27. A. H. Jaffe and N. Kaiser, "Likelihood Analysis of Large-Scale Flows," *Astrophysical Journal* 455 (1995): 26, doi:10.1086/176551.

28. J. R. Bond, A. H. Jaffe, and L. Knox, "Estimating the Power Spectrum of the Cos-mic Microwave Background," *Physical Review D* 57 (1998): 2117–2137, doi:10.1103/physrevd.57.2117; J. R. Bond, A. H. Jaffe, and L. Knox, "Radical Com-pression of Cosmic Microwave Background Data," *Astrophysical Journal* 533 (2000): 19, 37, doi:10.1086/308625.

29. J. R. Bond and G. Efstathiou, "The Statistics of Cosmic Background Radiation Fluctuations," *Monthly Notices of the Royal Astronomical Society* 226, 3 (1987): 655–687, doi:10.1093/mnras/226.3.655.

30. P. de Bernardis et al., "A Flat Universe from High-Resolution Maps of the Cosmic Microwave Background Radiation," *Nature* 404, 6781 (2000): 955–959, doi:10.1038/35010035.

31. Wilkinson Microwave Anisotropy Probe, https://wmap.gsfc.nasa.gov.

32. Data from NASA's LAMBDA website, https://lambda.gsfc.nasa.gov/. Plotted using the Healpix software (Gorski et al., "HEALPix.").

33. W. L. Freedman, "Final Results from the Hubble Space Telescope Key Project to Measure the Hubble Constant," *Astrophysical Journal* 553, 1 (2001): 47–72, doi:10.1086/320638.
34. "Planck," European Space Agency, https://www.esa.int/planck; Planck Collaboration, "Planck 2018 Results, I. Overview, and the Cosmological Legacy of Planck," *Astronomy & Astrophysics* 641 (2020): A1, doi:10.1051/0004–6361/201833880.
35. Data from the Planck Legacy Archive, https://wiki.cosmos.esa.int/planck-legacy-archive. Plotted using the Healpix software (Gorski et al., "HEALPix.").

Chapter 11. The Inflationary Universe

1. V. F. Mukhanov and G. V. Chibisov, "Quantum Fluctuation and 'Nonsingular' Universe," *JETP Letters* 33 (1981): 532–535; S. W. Hawking, "The Development of Irregularities in a Single Bubble Inflationary Universe," *Physics Letters B* 115, 4 (1982): 295–297, doi:10.1016/0370–2693(82)90373–2; A. Starobinsky, "Dynamics of Phase Transition in the New Inflationary Universe Scenario and Generation of Perturbations," *Physics Letters B* 117, 3–4 (1982): 175–178, doi:10.1016/0370–2693(82)90541-X; A. H. Guth, "Fluctuations in the New Inflationary Universe," *Physical Review Letters* 49, 15 (1982): 1110–1113, doi:10.1103/PhysRevLett.49.1110; J. Bardeen, P. Steinhardt, and M. Turner, "Spontaneous Creation of Almost Scale-Free Density Perturbations in an Inflationary Universe," *Physical Review D* 28, 4 (1983): 679–693, doi:10.1103/PhysRevD.28.679.
2. Data from the Planck Legacy Archive, http://pla.esac.esa.int/pla. doi.org/10.5270/esa-gb3sw1a.
3. J. Levin, *Black Hole Blues and Other Songs from Outer Space* (Knopf, 2016).
4. BICEP2 Collaboration, "Detection of B-Mode Polarization at Degree Angular Scales by BICEP2," *Physical Review Letters* 112 (2014): 241101, doi:10.1103/PhysRevLett.112.241101.
5. Planck Collaboration, "Planck 2013 Results. XVI. Cosmological Parameters," *Astronomy & Astrophysics* 571 (2014), doi:10.1051/0004–6361/201321591.
6. Planck Collaboration, "Planck Intermediate Results. XXX. The Angular Power Spectrum of Polarized Dust Emission at Intermediate and High Galactic Latitudes," *Astronomy & Astrophysics* 586, 133 (2016): A133, doi:10.1051/0004–6361/201425034; BICEP2/Keck and Planck Collaborations, "Joint Analysis of BICEP2/Keck Array and Planck Data," *Physical Review Letters* 114 (2015): 101301.
7. B. Keating, *Losing the Nobel Prize* (W. W. Norton, 2018).

Chapter 12. Tension or Concordance?

1. B. Chaboyer, "Absolute Ages of Globular Clusters and the Age of the Universe," *Astrophysical Journal* 444 (1995): L9.

2. A. H. Jaffe, "H_0 and Odds on Cosmology," *Astrophysical Journal* 471 (1996): 1, doi:10.1086/177950.

3. J. P. Ostriker and P. J. Steinhardt, "The Observational Case for a Low Density Universe with a Nonzero Cosmological Constant," *Nature* 377 (1995): 600–602, doi:10.1038/377600a0; L. M. Krauss and M. S. Turner, "The Cosmological Constant Is Back," *General Relativity and Gravitation* 27 (1995): 1137–1144, doi:10.1007/BF02108229; N. A. Bahcall et al., "The Cosmic Triangle: Revealing the State of the Universe," *Science* 284 (1999): 5419, doi:10.1126/science .284.5419.1481.

4. A. H. Jaffe et al., "Cosmology from MAXIMA-1, BOOMERANG, and COBE DMR Cosmic Microwave Background Observations," *Physical Review Letters* 86, 16 (2001): 3475–3479, doi: 10.1103/PhysRevLett.86.3475.

5. Planck Collaboration, "Planck 2015 results. XIII. Cosmological Parameters," *Astronomy & Astrophysics* 594 (2016): A13, doi: 10.1051/0004–6361/201525830.

6. R. R. Caldwell, R. Dave, and P. J. Steinhardt, *Physical Review Letters* 80 (1998): 1582.

7. A. G. Riess et al., "A Comprehensive Measurement of the Local Value of the Hubble Constant with 1 km s-1 Mpc-1 Uncertainty from the Hubble Space Telescope and the SH0ES Team," *Astrophysical Journal Letters* 934 (2022): L7.

8. M. Rowan-Robinson, *The Cosmological Distance Ladder: Distance and Time in the Universe* (W. H. Freeman, 1985).

9. Planck Collaboration, "Planck 2018 Results—VI. Cosmological Parameters," *Astronomy & Astrophysics* 641 (2020): A6, doi: 10.1051/0004–6361/201833910.

10. V. Poulin et al., "Early Dark Energy Can Resolve the Hubble Tension," *Physical Review Letters* 122, 221301 (2019).

11. W. L. Freedman and B. F. Madore, "Progress in Direct Measurements of the Hubble Constant," arXiv:2309.05618 (2023); A. J. Lee et al., "The Chicago-Carnegie Hubble Program: The JWST J-region Asymptotic Giant Branch (JAGB) Extragalactic Distance Scale," arXiv:2408.03474 (2024).

12. C. J. Copi et al., "Large-Angle Anomalies in the CMB," *Advances in Astronomy* (2010): 847541, doi:10.1155/2010/847541.

13. J. Levin, *How the Universe Got Its Spots: Diary of a Finite Time in a Finite Space* (Princeton University Press, 2002).

14. A. Niarchou and A. H. Jaffe, "Imprints of Spherical Nontrivial Topologies on the Cosmic Microwave Background," *Physical Review Letters* 99 (2007): 81302, doi:10.1103/physrevlett.99.081302; Planck Collaboration, "Planck 2013 Results. XXVI. Background Geometry and Topology of the Universe," *Astronomy & Astrophysics* 571 (2014): A26, doi:10.1051/0004–6361/201321546; Planck Collaboration, "Planck 2015 Results. XVIII. Background Geometry and Topology of the Universe," *Astronomy & Astrophysics* 594 (2016): A18, doi:10.1051/0004–

6361/201525829; Y. Akrami et al., "Promise of Future Searches for Cosmic Topology," *Physical Review Letters* 132 (2024): 171501, doi:10.1103/physrevlett.132.171501.

15. E.g., I. Labbé et al., "A Population of Red Candidate Massive Galaxies ~600 Myr After the Big Bang," *Nature* 616 (2023): 266–269, doi:0.1038/s41586–023–05786–2.

Chapter 13. The Limits of Knowledge

1. T. M. Davis and C. H. Lineweaver, "Expanding Confusion: Common Misconceptions of Cosmological Horizons and the Superluminal Expansion of the Universe," *Publications of the Astronomical Society of Australia* 21, 1 (2004): 97–109, doi:10.1071/AS03040.

2. A. Albrecht and P. J. Steinhardt, "Cosmology for Grand Unified Theories with Radiatively Induced Symmetry Breaking," *Physical Review Letters* 48, 17 (1982): 1220–1223, doi:10.1103/PhysRevLett.48.1220; A. D. Linde, "Eternally Existing Self-Reproducing Chaotic Inflationary Universe," *Physics Letters B* 175, 4 (1986): 395–400, doi:10.1016/0370–2693(86)90611–8.

3. W. James, *Is Life Worth Living?* (1896). In contrast, "universe" comes from the Latin "universum," with roughly the same all-encompassing meaning as the term holds in English today.

4. M. Peters, "The Geeky, Cosmic, Philosophical History of "Multiverse," Vocabulary.com, https://www.vocabulary.com/articles/wordroutes/the-geeky-cosmic-philosophical-history-of-multiverse/.

5. M. Tegmark, *Our Mathematical Universe: My Quest for the Ultimate Nature of Reality* (Knopf, 2014).

6. B. Carter, "Large Number Coincidences and the Anthropic Principle in Cosmology," *IAU Symposium 63: Confrontation of Cosmological Theories with Observational Data* 63 (1974): 291–298, doi:10.1017/S0074180900235638, republished online by Cambridge University Press (February 7, 2017).

7. S. Weinberg, "Anthropic Bound on the Cosmological Constant," *Physical Review Letters* 59, 22 (1987): 2607–2610, doi:10.1103/PhysRevLett.59.2607.

8. J. D. Barrow and F. J. Tipler, *The Anthropic Cosmological Principle* (Oxford University Press, 1986); P. Davies, *The Goldilocks Enigma* (Allen Lane, 2006). See also my review of Davies: A. H. Jaffe, "Seeking Anthropic Answers," *Physics World* (November 1, 2006), https://physicsworld.com/a/seeking-anthropic-answers/.

9. E. Kolb and M. Turner, *The Early Universe* (Addison-Wesley, 1990).

10. I. Hacking, "The Inverse Gambler's Fallacy: The Argument from Design. The Anthropic Principle Applied to Wheeler Universes," *Mind* 96, 383 (1987): 331–340,

doi:10.1093/mind/XCVI.383.331; R. M. Neal, "Puzzles of Anthropic Reasoning Resolved Using Full Non-indexical Conditioning," arXiv:math/0608592 (2006).

11. A. Albrecht and L. Sorbo, "Can the Universe Afford Inflation?," *Physical Review D* 70, 6 (2004): 063528, doi:10.1103/PhysRevD.70.063528; A. Linde, "Sinks in the Landscape, Boltzmann Brains and the Cosmological Constant Problem," *Journal of Cosmology and Astroparticle Physics* (2007): 1, 22; A. Aguirre, S. M. Carroll, and M. C. Johnson, "Out of Equilibrium: Understanding Cosmological Evolution to Lower-Entropy States," *Journal of Cosmology and Astroparticle Physics* (2012): 2, 24, doi:10.1088/1475–7516/2012/02/024.

12. D. Lewis, *On the Plurality of Worlds* (Blackwell, 1986).

13. Tegmark, *Our Mathematical Universe*.

INDEX

Page numbers in italics *refer to figures.*